Pollinators in CRISIS

How we can ALL give them a helping hand

by
Michael Fogden

Photographs by Michael and Patricia Fogden

piscespublications

Published 2022 by Pisces Publications

First published 2022.

British-Library-in-Publication Data
A catalogue record for this book is available from the British Library.

ISBN 978-1-913994-04-4

Designed and published by Pisces Publications

Pisces Publications is the imprint of NatureBureau,
2C The Votec Centre, Hambridge Lane, Newbury, Berkshire RG14 5TN
www.naturebureau.co.uk

Printed and bound in the UK by Gomer Press Ltd

All the paper used is FSC mixed source certified. All inks are colour-fast, vegetable derived (no petrochemicals involved). Each page is dried under UV in seconds, saving energy on hours of costly heating. All printing is digital, no metal plates involved. Production is by Gomer Press in South Wales, whose environmental credentials can be inspected at http://gomerprinting.co.uk/environment/ They are accredited to the Green Dragon Environmental Standard Level 2 https://www. greenbusinesscentre.org.uk/green-dragon-environmental-standard

Front cover Tree Bumblebee on Hollyhock flower
Back cover Wildflower meadow, Wendlebury Meads, Oxfordshire [PC]

Visit our bookshop to see our full range of books
www.naturebureau.co.uk/bookshop/

Contents

Foreword

In recent decades there have been significant, one might even say potentially catastrophic declines in the number of insects in the countryside, not just in the UK but worldwide. Many insects are pollinators, so *Pollinators in Crisis* with its subtitle *How we can ALL give them a helping hand* is a timely book to help educate, at a time when crops play a huge part in global food production; further declines in pollinators could threaten population health.

I have read various books and scientific papers on the subject and at last, governments and policy makers are taking notice of concerns. This different work is a passionate take on the subject by a zoologist who has worked on various research projects worldwide, photographing and writing about wildlife. One cannot help to notice, even in the first few pages, amazing, vibrant colours and the high standard is maintained throughout this book. Understandably, some of the common pollinators feature in photos several times.

This work is up to date and well designed, with interesting observations from around the world, but mainly the UK. The chapters are well thought out and neatly divided into sections, explaining why insects are declining. The book opens on a positive note in **Part 1 Flowers and people**, *"Flowers make us smile and feel happy."* Further short sections 'Flowers in human culture' and 'Flowers and food' then lead the reader to **Part 2 The countryside, agriculture and pollinators**. Sections here include global warming, updated to include reference to COP26. **Part 3 Saving pollinators** and **Part 4 The natural history of pollination** provide much useful background, and emphasise the type of information, for those who want to help reverse the pollinator crisis in the UK. Gardeners will find this invaluable. Examples are given of how plants reward pollinating insects and the wide range of pollinators, not just bees. **Part 5 Flowers for pollinators** is a large chunk of the book (90 pages) and provides a superb, illustrated account of 100 flowers that grow in the author's own garden in Norfolk, many illustrated by associations with different pollinator species. So those who are keen to assist can ALL provide practical help and enjoy a visual feast of some of the author's recommended flowers to ensure a range of plants to attract insects in all seasons.

This book will surely be treasured by those with an interest in or wanting to learn more about flowers, insects and pollination and gardeners looking for inspiration. The more we know the better; I have been studying insects for more than 50 years, but still found some information new to me. Readers will gain a good understanding of the several reasons why pollinators are in crisis, what has been done recently and what must be done to help reverse the declines. The dazzling photographs are quite simply superb.

Paul D. Brock
Scientific Associate, Natural History Museum
November 2021

Preface

This book is about flowers, how they are pollinated, and the pollination crisis that is impacting the UK and much of the world. It is divided into five parts.

Part 1 Flowers and people describes the ways in which flowers are involved in the lives of people round the world. Flowers have long played an important role in our cultural activities and the end products of flowers—their fruits and seeds—are vitally important as food. The dawn of farming, 10–12,000 years ago, resulted in enormous changes to human societies. Farming allowed hunter-gatherer societies to settle and grow crops. Many crops have flowers that need pollinators—about 75% of the world's food crops rely at least in part on pollination by insects and other animals and together make up over one third of global crop production. The economic value of pollinator services to agriculture is well documented and amounts to between £430 million and £690 million per year in the UK alone.

Part 2 The countryside, agriculture and pollinators is concerned with modern intensive agriculture which involves the continuing destruction of natural habitats and the excessive use of herbicides, insecticides and other agricultural chemicals. This has resulted in alarming declines of bees and other pollinators. Most farmland in the UK, and elsewhere in the world, has become a hostile environment for pollinators. If allowed to continue, the current pollinator crisis is likely to have dire consequences for agriculture and food. Global warming is also discussed in Part 2. It is an important factor contributing to pollinator declines, particularly because it often results in a temporal mismatch—a lack of synchronisation between the timing of flowering and the seasonality of appropriate pollinators.

Part 3 Saving pollinators discusses ways to reverse pollinator declines. Stimulated by the publicity that the pollinator crisis has provoked, some progress is being made to improve matters. Efforts are being made to reduce the impact of agricultural chemicals and to link existing areas of natural habitat with flower-rich corridors, such as motorway verges, railway lines and field verges. Gardens can be an important part of the solution by becoming flower-rich oases, providing nectar and pollen, breeding sites and green stepping-stones that link scattered fragments of other useful habitat.

Part 4 The natural history of pollination is concerned with the natural history of flowers and pollination. If the ongoing pollinator crisis in the UK is to be reversed, gardeners and anyone else interested in helping should know enough about pollination biology to ensure that a useful selection of flowers are in place to create year-round, flower-rich refuges where pollinators can prosper. Some flowers are pollinated by wind, and a few by water, but most have evolved to send signals to insects and other animal pollinators to attract and reward them (or deceive them) into carrying their pollen from flower to flower. Different flowers have different suites of

attractive characters, known as 'pollination syndromes', which match the sensory capabilities of their pollinators.

 **Part 5 Flowers for pollinators** recommends about 100 native and garden flowers (in 35 plant families) that are good for pollinators and describes features of their natural history that are specifically related to pollination or are of general interest. Many of our recommendations are beautiful flowers that are commonly found in gardens. Many others are so-called 'weeds', most of them exceptionally good for pollinators and (at least in our opinion) often beautiful in their own right.

Our book is mainly about the UK's native, naturalised and common garden flowers and their pollinators but we have mentioned some foreign material to provide world context. Whenever possible, we have used well-established English common names for plants and animals. For insect pollinators, well-established common names exist mainly for bees, butterflies, moths and a few flies and beetles. We have, therefore, used scientific names for some insects and whenever we feel extra clarity is needed. Our main sources for names (both common and scientific) are the field guides listed in 'Further reading'. The latter also includes important scientific research papers and other source material. We have avoided technical botanical terms as much as possible but a good number are unavoidable. Definitions for those used can be found in the Glossary.

Michael Fogden
November 2021

Acknowledgements

Special thanks are due to Tricia—constant companion, director of photography and chief gardener. Tricia lovingly tends our weedy garden. She also read the manuscript several times and, as always, said exactly what she thought of it. Tricia also took a good number of the photographs of flowers and their visiting pollinators.

We are also indebted to all those scientists who have worked on pollination, pollinators and the pollinator crisis. Several of the most important or useful studies and sources are listed in 'Further reading' on pp. 172–173.

Both Tricia and I are most grateful to friends who invited us into their gardens to take photographs; gave us many interesting plants, including exceptional weeds; and also provided excellent and much appreciated hospitality. We are very grateful to Chris and Annie Abrams, Leo and Angela Batten, Roger and Margo Brownsword, Nick and Frankie Owens, Richard and Julia Porter, David and Lindsey Wall, and last, but by no means least, Barbara Woodcock and the late Martin Woodcock. We are especially indebted to Nick Owens for his expert identifications of solitary bees.

A number of photographs were supplied by others without charge and we are grateful to Paul Brock [PB], Peter Creed [PC], Peter Eeles [PE], Ian Duncan [ID] and Nick Owens [NO] for the use of several of their excellent images. Their intials are placed in square brackets at the end of the captions. Finally, we are grateful for the enthusiastic support of Peter Creed, Creative Director of NatureBureau and all involved at Pisces Publications.

PART 1
Flowers and people

Field of Corn Poppies *Papaver rhoeas* [PC]

CHAPTER 1
Flowers in human culture

Flowers make us smile and feel happy. A love of flowers is almost universal and it is no surprise that flowers have long played an important role in the cultural and artistic lives of people around the world. Flowers with special symbolic importance include the rose in western countries; lotus flowers in Buddhist and Hindu cultures in India; peonies, chrysanthemums and orchids in China; and cherry blossom and chrysanthemums in Japan. Flowers have long been a popular choice for sending messages of love to sweethearts, wives and mothers. In fact, the two most important flower days in the western world are Valentine's Day and Mothers' Day, the former being a day, according to the National Retail Federation in the USA, when over 250 million roses are produced for the US market and US consumers spend $1.9 billion on flowers. UK consumers are said to spend over £100 million on flowers for Valentine's Day. Flowers are also used to honour death and sacrifice. Over much of the world, bouquets are placed on graves and at the sites of tragedies—flower-decked, roadside shrines are often seen at the sites of fatal traffic accidents or terrorist incidents. Remembrance Day, or 'Poppy Day' as it is more commonly called in the United Kingdom, commemorates the millions who lost their lives in the First World War, during which blood-red poppies bloomed profusely amongst the guns, shells, death and destruction of the Flanders battlefields and have come to represent the blood spilled in the war and the lives lost.

Cherry blossom in Japan [Lê Tuấn Hùng (pxhere.com)]

Flowers have also inspired artists and photographers. The best-known flower paintings in the western world are probably those painted by French impressionist Claude Monet, who passed the last 30 years of his life painting water-lilies in his garden at Giverney, and Vincent Van Gogh's many paintings of sunflowers and irises. Other notable flower images include paintings by Édouard Manet, Jan Brueghel the Elder, Georgia O'Keeffe and Andy Warhol, as well as photographs by Edward Weston, Imogen Cunningham and Robert Mapplethorpe. Flower imagery is equally popular in Chinese, Japanese and Indian cultures. In China, for example, peonies, chrysanthemums, hibiscus and other flowers feature in 'bird-and-flower' paintings and on porcelain, dating from at least the Sung dynasty (960–1279). Flower images are also common motifs on antique Persian and other Middle Eastern and Oriental rugs.

Water-lilies by Claude Monet
[Musée Marmottan-Monet, Paris (public domain)]

Gardens are a notable expression of mankind's love of flowers. Ornamental gardens first appeared sometime after the rise of agriculture when settled communities replaced many hunter-gatherer societies. Food surpluses led to a division of labour and a demand for workers with specialist skills—metal workers making weapons, tools and jewellery, potters, stone masons, warriors, musicians, priests and probably gardeners. Specialisation led to a less egalitarian society and the emergence of privileged, upper and ruling classes with leisure time and a taste for comfort and luxury. Ornamental gardens were luxuries that existed in Egypt and Mesopotamia at least 5–6,000 years ago and 2–3,000 years ago in China. In Egypt and Mesopotamia, gardens were retreats from fierce, desert heat. They were cool, shady spaces, with trees, ponds with water-lilies, and flower beds. Later, ornamental gardens spread to western Europe, brought by the Romans from Egypt and by the spread of Islam into Spain. It was the Romans who introduced plums, walnuts, roses and parsley to British gardens.

In later centuries, exotic plants arrived in Britain from all over the world. Sir Francis Drake brought potatoes and tobacco from America. Other early explorers of the Americas introduced pumpkins, runner beans, sweet corn, tomatoes and many North American trees. British gardening received a massive boost in the Victorian era, when European botanical adventurers went plant hunting around the world. China, Japan, and the Himalayas were the most productive areas, the origin of rhododendrons, azaleas, camellias, hollyhocks, buddleia, flowering cherries, primulas and a host of other species that soon became garden favourites.

The Victorian era also saw the phenomenon that became known as 'orchidelerium' or 'orchid madness'. Wealthy collectors dispatched orchid hunters around the world to search for new, spectacular species. Many collectors failed to return, having reputedly been killed by hostile natives, eaten by cannibals or tigers, or drowned in rapids. New species were sold at auction and fetched high prices. In 1890, £1,500 was paid for a single orchid, equivalent to around £100,000 today. More recently, in 2005, the Shenzhen Nongke Orchid—a not particularly spectacular *Cymbidium* variety bred in China—was sold at auction for 1.68 million Yuan, equivalent then to a staggering £160,000.

Today flowers have great economic importance. There is a lucrative and expanding global market for cut flowers, house plants and garden plants. In the 1950s the global flower trade was worth less than $3 billion/year but grew to over $100 billion by 2017. The Dutch have been leaders of the global flower trade since the beginning of the 20th century. Daily auctions are held at the Aalsmeer Auction House and flowers are then despatched around the world from nearby Schipol Airport. The auctions take place in a building covering over 50 hectares, the world's largest by area, and involves the sale of over 20 million flowers daily. But markets are shifting. A new, hi-tech auction house—the Dubai Flower Centre—opened in 2005 and flower production is expanding rapidly at high altitudes in developing countries where the climate is favourable all year round and labour costs low. Nowadays, cut flowers are flown great distances from tropical countries, notably Colombia, Ecuador, Costa Rica, Kenya, and Uganda, to major markets in the USA, Europe, and Japan.

A few flowers have more varied and specialised uses. Lavender is a good example and is used in many ways—as garden flowers; to make bundles of fragrant dried flowers; in sachets, soap and lotions; and especially to produce an essential oil used in aromatherapy and perfumes. Lavender is known to have been used in ancient Egypt in the process of making mummies. Later it was used as a scented bath additive in ancient Persia, Greece and Rome. Nowadays, it is used in aromatherapy to treat numerous ailments, including anxiety, insomnia, depression, headaches, toothache, acne and even hair loss! Lavender is grown commercially for lavender oil in many countries, including Provence in France, England, Italy, Spain, South Africa, California and Australia.

Tulips growing in the Netherlands for cut flowers [pxhere.com]

Lavender crop in Provence, France [pxhere.com]

CHAPTER 2
Flowers and food

The **final bounty** of flowers—that is their fruits and seeds—have long been important items in the human diet. The earliest humans collected berries, nuts and seeds from the wild and hunter-gatherer tribes in Africa, Amazonia and elsewhere still do so. The dawn of farming, around 10–12,000 years ago, and the population explosion that followed, resulted in enormous change. People no longer had to keep moving to find food. Farming enabled them to live in settled communities and build permanent homes. This, in turn, allowed the cultivation of crops and the growth of orchards, the latter demanding decades-long commitments of time. Settled agricultural communities could produce food surpluses and support sufficient numbers of non-cultivating specialists to give rise to the first civilizations. New farming techniques were developed and valuable food plants were improved by selective breeding for useful qualities and carried to new areas, often well away from their original natural range. Many important crops, including rice, wheat, maize, potatoes, coffee, grapes, apples, oranges, bananas, strawberries and many others, are now cultivated far beyond their original, native range, restricted only by their need for appropriate growing conditions.

The most important staple crops consumed throughout the world today are the three cereals—maize, rice and wheat—all of them grasses with flowers that are wind-pollinated rather than pollinated by insects. Domestication of these cereals and other staples is thought to have started at the end of the last ice age, about 12,000 years ago, a time when the onset of more favourable temperatures saw the rise of agriculture in several parts of the world.

Wild rice was being collected in China at least 12,000 years ago and has been farmed for 8,000 years. From China, rice cultivation spread westwards, reaching India 5,000 years ago and the Mediterranean basin around 344–324 BC. Wheat, first domesticated in the Fertile Crescent (the region stretching in an arc from the Tigris and Euphrates rivers to the upper Nile), is another crop that has been grown for about 12,000 years. Numerous varieties have been developed since, some of the most important being Common Wheat *Triticum aestivum* used to make bread; Durum Wheat *Triticum durum* used for spaghetti, macaroni and other pastas; and Club Wheat *Triticum compactum* used for pastries, cakes and cookies. The identity of the wild ancestor of Maize (often called corn) remained a mystery until the 1930s. Only then was it established that the ancestor was a type of grass native to Mexico called 'teosinte' *Zea mays* ssp. *parviglumis*. It looks nothing like the Maize that is grown today but is nevertheless similar in its DNA and can cross-breed with modern maize varieties. It was first domesticated about 9,000 years ago in the Balsas River drainage in southern Mexico.

Rice paddy fields, Bali, Indonesia

An important non-grain staple—the Potato *Solanum tuberosum*—with flowers that are mostly pollinated by bees, is thought to have been first cultivated in the Andes of South America (in the Titicaca basin of southern Peru) at least 5,400 years ago. Also in South America, recent research published in the journal *Nature* has revealed that crops were also being grown in Amazonia, in what is now northern Bolivia, several thousand years ago. Both Manioc *Manihot esculenta*, also known as Cassava or Yuca, and squash were being cultivated in the area well over 10,000 years ago and maize 6,850 years ago.

Two other important commodities are obtained from the seeds of cocoa and coffee trees, both of which have flowers that are insect pollinated. Cocoa (or chocolate) is made from the fermented, dried and roasted seeds (or beans) of *Theobroma cacao*, a tree species from the rainforests that extend from southern Mexico to the Amazon Basin. Cocoa was first exploited more than 5,000 years ago in the Amazon rainforests of Ecuador. Later it was carried north to Mexico and Central America, where it was cultivated by the Aztecs and Mayans. Later still, in the mid-16th century, cocoa was carried to Spain by Spanish conquistadors and from there, over the next 100 years, it spread to France, England and elsewhere in western Europe. The history of coffee, a native of northern Africa, is less ancient than that of cocoa. Its first use is thought to date back to the 10th century in Ethiopia, though the first substantiated record of coffee drinking is from Sufi monasteries in Yemen in the mid-15th century. Coffee subsequently reached Europe in the 17th century. Its arrival in Venice provoked fear in some who called it the "bitter invention of Satan". Local clergy condemned it and asked Pope Clement VIII to adjudicate on its use. The Pope tasted coffee for himself and clearly liked it. He gave his papal approval and the rest is history—coffee houses quickly became focal points for social activity and intellectual debate in major cities throughout Europe. By the mid-17th century there were more than 300 coffee houses in London, some of which are said to have evolved into the London Stock Exchange, Lloyds of London, and the auction houses of Sotheby's and Christie's. At the present time, *Coffea arabica*, native to Ethiopia, is the source of gourmet coffees which together amount to about 70% of coffee production worldwide, most of it now grown in Brazil. The other 30% is sourced from a different species—*Coffea canephora*—widely called "robusta coffee", a species that is more hardy than *C. arabica* and native to areas further south in sub-Saharan Africa. After Brazil, the most important coffee producing countries are Vietnam, Colombia, Costa Rica,

Cocoa pod, Costa Rica

Coffee tree flowering [pxhere.com]

Indonesia, India and Ethiopia. And it may come as a surprise to many to learn that coffee is, after oil, the second most important traded commodity in the world, ahead even of natural gas, gold and wheat. Coffee provides employment for more than 25 million people.

Several of the world's most popular fruits have also been grown for several thousands of years. Archaeological evidence demonstrates that figs were being farmed in the Jordan Valley at least 11,000 years ago, making them the first non-cereal crop to be cultivated anywhere in the world. Later, figs were introduced throughout the Mediterranean region by the Greeks and Romans. Dates are another ancient crop. Dates have been a cultivated staple food for over 9,000 years and probably originated in Mesopotamia. Dates are naturally adapted to be wind-pollinated but cultivated dates have always been hand-pollinated, and still are in commercial orchards. Grapes were first grown to make wine over 6,000 years ago in Armenia, a practice perhaps discovered by lucky accident because wild yeast occurs naturally on the skins of grapes, eventually resulting in fermentation and the formation of alcohol. Other important fruits that have been cultivated for thousands of years, include olives, originally from the Middle East; oranges, peaches and apricots from China; apples from central Asia; pomegranates from Persia; and mangos, oranges and bananas from India.

Wheat, rice, maize and other cereals are wind-pollinated but around 76% of today's globally important crops, including coffee, cocoa (chocolate), figs, melons, apples, pears, avocados, almonds, strawberries, blueberries and most beans, depend to a greater or lesser extent on pollination services provided free of charge by bees or other insects. As well as fruits and vegetables, other relevant crops that depend on pollination by insects include sunflowers, rapeseed, canola and oil palm, grown as sources of vegetable oils and bio fuels; cotton for the manufacture of textiles; and alfalfa and clover, grown as pasture to be grazed by livestock. Some of these plants are capable of self-pollination but cross-pollination is almost always preferable because it results in greatly improved crop yields and better quality produce.

Rapeseed crop, Essex [Anik Messier]

The economic value of pollinator services is well documented. According to the first global assessment of pollinators, a two-year study carried out by the Intergovernmental Science-Policy Platform on Biodiversity and Ecosystem Services (IPBES), the global contribution of pollinators to agricultural production, through increased yields and quality, is worth between US$235 billion and US$577 billion per year. Another survey estimated the gross economic value of the 105 most widely planted, insect-pollinated crops to be more than US$800 billion per year. Equivalent figures for the UK alone are between £430 million and £690 million. And, beyond their monetary value, insect pollinators make a priceless contribution to biodiversity by pollinating the plants that provide the fruits and seeds that are eaten by a host of mammals, birds and other animals. It must also be emphasised that these pollination services are provided by hundreds of species of wild bumblebees, solitary bees, hoverflies, butterflies and other insect pollinators, not just by domesticated Honey Bees *Apis mellifera*. The Honey Bee may be an important species for crop pollination, mainly because hives can be transported to take care of monocultures. It is also relevant that the average annual production of honey worldwide is over 1.8 million metric tons and provides an important source of income in many rural communities. However, there is mounting evidence that managed Honey Bees do little more than supplement the free services of wild pollinators, no matter how many hives of Honey Bees are available (see below p. 11). The survival of the numerous wild pollinators is vital if the productivity of insect-pollinated crops and food security is to be maintained. In general, Honey Bees are less important.

PART 2
The countryside, agriculture and pollinators

View of the countryside and the Thames near Goring from Hartslock nature reserve, Oxfordshire [PC]

CHAPTER 3
The pollinator crisis

Over recent years, worrying declines of Honey Bees (and other pollinators, including wild bees, beetles, butterflies and moths), particularly in Europe and North America, have received a lot of publicity in the media—in the UK barely a day goes by without news items about bees on television, in newspapers and online. It is clear that bees and other important pollinators, especially hoverflies, are in trouble, mostly from habitat loss and the impact of intensive agriculture. The human population of the world has almost doubled in the past 50 years and average calorie consumption per person has increased by 30%. The ever increasing demand for land for food production has resulted in serious losses of forest, grassland and other natural habitats where bees can forage and breed, as well as exposure to more and more toxic insecticides, herbicides and other agricultural chemicals.

A recent study, published in *Nature Communications*, looked at population trends in 353 species of wild bees and hoverflies in the UK over a period of 33 years starting in 1980. A third of the species suffered declines, with every square kilometre of land losing an average of 11 species of bees and hoverflies during the 33 years of the study. Obviously, local diversity has been badly affected.

Similar declines in insect diversity are being recorded worldwide. A recent review of 73 previous studies, published in *Biological Conservation* revealed "*dramatic rates of decline that may lead to the extinction of 40% of the world's insect species over the next few decades*". According to lead author Dr Francisco Sánchez-Bayo, the driving factors of the declines are: habitat loss and conversion to intensive agriculture and urbanisation; pollution due to agricultural pesticides, herbicides and fertilisers; pathogens and introduced species; and climate change, the latter being particularly important in tropical regions. The authors also note that insect declines have important knock-on effects up the food chain, especially on birds. Commenting on the study, Dave Goulson, Professor of Biology at the University of Sussex, said there were things that people could do, such as making their gardens more insect friendly, not using pesticides and buying organic food. He added that if huge numbers of insects disappear, they will be replaced but it will take a long, long time. "*So give it a million years and I've no doubt there will be a whole diversity of new creatures that will have popped up to replace the ones wiped out in the 20th and 21st centuries. Not much consolation for our children, I'm afraid.*"

The recent declines are a reason for immediate concern because bees, both wild bees and domesticated Honey Bees, hoverflies and other insects, pollinate around 75% of the crops in the world and play a crucial role in food production. Since 1985, Honey Bees have declined by about 25% in much of Europe and over 50% in the UK. Going back further, there were about 250,000 viable Honey Bee colonies in the UK in the 1950s, compared with fewer than 100,000 today. Parts of the world that have also suffered major losses of Honey Bees include the USA, China, Japan, and the Nile valley of Egypt. The losses have become so bad that wild bees are now more important that domesticated Honey Bees as pollinators of crops around the world. A recent, large and important international survey of insect pollinators found that just 2% of wild bee species now account for as much as 80% of crop pollination in many parts of the world.

The many reasons for the demise of Honey Bee colonies, known as 'colony collapse disorder', involve pathogens and parasites, the overuse of pesticides and herbicides, poor nutrition, the loss of suitable habitat, climate change, or several of these factors in combination. The parasitic mite, *Varroa destructor*, is a significant menace. Originally, *Varroa destructor* was native to Asia and parasitic on the Asian Honey Bee. It is now found in Honey Bee populations throughout much of the world, with the notable exception of Australia. The mites feed on the blood of both Honey Bees and their larvae and, in doing so, are liable to spread harmful diseases. Bee colonies infected by *Varroa* mites are usually killed off within two or three years.

Unfortunately, most of the publicity about losses of bees has concentrated far too much on Honey Bees. Managed Honey Bees are essential for a few crops, but only because the crops are monocultures so large that there are no flowers for wild pollinators or, in the case of monocultures that do have useful flowers, they are available for only two or three weeks each year. Hence there are almost no wild bees or other pollinators. For appropriate crops, Honey Bees can fill the gap because their hives are easily moved to locations near monocultures and provide a temporary pollination service (but see pp. 12–13. California's vast Almond groves are an extreme example of the monoculture scenario. For pollination, the Almond blossom is now entirely dependent on domesticated Honey Bees that have to be transported to California from all around the USA—a total of 1.7 million hives are involved, about 85% of all the hives available in the USA. Once the Almond groves have been serviced, the Honey Bees are moved on again and again to take care of other crops—cherries, citrus fruits, blueberries and so on.

The Almond groves are the source of another problem. California is prone to drought and each Almond produced is said to need a gallon of water. In total, therefore, the Almond crop uses as much water in a single year as homes and businesses in Los Angeles use in three years. Given that water tables are dropping in many parts of California, this is a matter of serious concern.

Though Honey Bees are obviously crucial for California's Almonds, for most crops in the UK and elsewhere there is mounting evidence that losses of wild bees are far more serious than the losses of domesticated Honey Bees. In an article in *Wired* magazine (the self-styled *Rolling Stone* of technology) science writer Gwen Pearson told her American audience that we are "*worrying about the wrong bees*", that Honey Bees are a "*globally distributed, domesticated animal*" and "*not remotely threatened with extinction*". She added "*The bees you should be concerned about are the 3,999 other bee species living in North America, most of which are solitary, stingless, ground-nesting bees you've never heard of. Incredible losses in native bee diversity are already happening.*"

Almond blossom, California

Or, in some areas, happened years ago. Research by American scientists looked at data collected in the Illinois countryside from 1888 until 1891 and compared it with similar data collected 80 and 120 years later, elapsed years during which the forested countryside of rural Illinois was reduced to just small isolated forest fragments. The results of the study showed that more than 50% of the original 100+ species of wild bees had become locally extinct and that the surviving species were no longer capable of adequately pollinating all the local plants.

There is also other research that shows that it is the losses of wild bees, rather than managed Honey Bees, that is most important. Research carried out by 50 agricultural scientists on crops growing in different parts of the world, revealed that overall wild bees were twice as effective as Honey Bees in pollinating the 41 different crops that were studied, which included almonds, cherries, coffee, mangos, watermelons, tomatoes, strawberries, blueberries, rapeseed, Red Clover, and others. The study also demonstrated that crops are less productive if wild bees are absent, even if an excess of Honey Bees are present and active. In the UK there is evidence that two-thirds of crop pollination is carried out by wild pollinators and only one-third by Honey Bees. By themselves, Honey Bees are nowhere near the equal of a whole community of wild pollinators. In any case, reliance on domesticated Honey Bees (as with the California Almond crop) is best avoided. A lone pollinator species is obviously more vulnerable to disease or other problems than an assemblage of many different pollinators.

It is also worrying that unnaturally high densities of managed Honey Bees may be actively doing harm to wild bees and other pollinators. Research shows that managed Honey Bees reduce numbers of wild pollinators in the vicinity of their hives in both natural habitats and crops. Perhaps worse, mass-flowering monocultures, such as Oil-seed Rape *Brassica napus* or field beans, attract very high densities of Honey Bees for just the few weeks they are flowering, after which the bees are forced to move to surrounding areas where they outcompete wild pollinators for scarce resources. The problem is exacerbated because the food supply of domesticated Honey Bees is often supplemented with sugar syrup or 'bee candy' when flowers are scarce.

Honey Bee at apple blossom

Supplemental feeding reduces mortality, resulting in unnaturally high population levels and yet more competition for wild bees. Honey Bees have also been linked to the spread of diseases to wild bees—a consequence made much worse by the transportation of Honey Bee hives around the countryside and from country to country.

Managed Honey Bees are no longer the only problem. Managed bumblebees have been used commercially for only a few years but are already causing major concerns. The domestication of bumblebees was first investigated back at the end of the 19th century because bumblebees, with their longer tongues, are able to pollinate a greater variety of flower types than shorter-tongued Honey Bees. However, it is only since the 1980s that domesticated bumblebees have been used commercially, first in the Netherlands and subsequently in numerous countries around the world, including the USA, Canada, Chile, Argentina, Japan and New Zealand. Bumblebees have proved to be effective pollinators of several crops, notably greenhouse tomatoes, a plant that previously had to be pollinated by hand or by using mechanical vibrators. Greenhouse grown tomatoes are now an important and profitable crop and are superior to tomatoes grown outdoors because they are allowed to ripen naturally on the vine. Tomatoes belong to a group of species that need to be buzz-pollinated—something that bumblebees do efficiently but Honey Bees do not do at all (see p. 47). Other crops that benefit from pollination by managed bumblebees include peppers, kiwifruit and many soft fruits, including raspberries, strawberries, cranberries and blueberries.

Unfortunately, recent research suggests that managed bumblebees may be implicated in declines of wild bumblebees in North America and probably elsewhere. In North America the fungal pathogen *Nosema bombi* is a 'key player' in the declines which occurred shortly after many commercial bumblebee operations collapsed due to *Nosema* infections. It is thought the *Nosema* pathogen must have escaped from commercial bumblebee colonies into wild populations. Elsewhere, the European Buff-tailed Bumblebee *Bombus terrestris* is now irreversibly established as a non-native species in several countries, including Japan, Chile and

Buff-tailed Bumblebee at thistle

Sunflower crop [epokrovsky/123rf]

Argentina, where it is a threat to native species through competition, hybridisation and the introduction of new non-native diseases. The spread of diseases that affect bees has increased enormously in recent years, exacerbated by the largely unregulated movement around the world of hundreds of thousands of Honey Bee and bumblebee colonies, along with their pathogens and parasites. If the pollinator crisis is not to get even worse, the export and import of commercial Honey Bees and bumblebees needs to be strictly controlled.

Problems with pollinators provide good examples of how our continuing disregard for the health of the environment threatens the future of agriculture. An especially significant example comes from the mountainous Maoxian region of China's Sichuan Province, where farmers (with thousands of helpers) were forced to pollinate every flower in their apple and pear orchards by hand, using pollen-loaded chicken feathers and paint brushes. Why? Because wild bees had been eradicated by intensive agriculture. Pollination by hand is feasible for high value crops in areas with plentiful cheap labour but it is too expensive in most circumstances and certainly not feasible in Europe or many other countries.

Bees and other pollinators are now facing yet another threat. Under the renewable fuel directive, the EU wants all EU countries to obtain at least 10% of their transport fuel from renewable sources, particularly biofuels, by 2020. So farmers are being encouraged to grow oil crops, such as sunflowers and Oil-seed Rape. Both these crops provide nectar and pollen, but they flower for just a few weeks at best, ensuring that there is little or nothing to sustain pollinators during the rest of the year. Scientists estimate that there is now a deficit of 13.4 million colonies of Honey Bees across Europe (amounting to about seven billion bees). Meanwhile, the even more important wild bees (bumblebees and solitary bees) continue to decline.

CHAPTER 4

The changing countryside

As is true elsewhere in the world, bees and other pollinators are declining in the UK for many different reasons, known and unknown, and often interrelated. However, there is little doubt that the countryside and farmland in the UK is so lacking in flowers nowadays that it has become a hostile environment for bees and other important pollinators. Only a few bees have become extinct in the UK but many have become locally extinct and now have a patchy distribution. As a result, bee diversity in most areas is now much lower than it used to be. Much the same is true for hoverflies and butterflies. And not just for pollinators. The countryside and farmland is also hostile for the predatory and parasitic insects, particularly wasps and flies, that help control agricultural pests, and the soil fauna that plays an essential role in the healthy functioning of agricultural environments. If nothing is done to reverse this state of affairs, declines in the health and populations of pollinators will continue to pose a major threat to the integrity of biodiversity, to food webs, and to human wellbeing.

A major reason for these problems—the loss of natural habitats—is particularly severe in the UK. By destroying forest and grasslands to create farmland, intensive agriculture has caused habitat for wild bees and other pollinators to diminish year after year, with a loss of suitable nesting sites as well as feeding resources. According to the *State of Nature Report 2016*, the UK is among the most nature-depleted countries in the world. As much as 75% of the UK's landscape is classified as agricultural and 40% as enclosed farmland (arable fields and improved grassland). With just 13% of its land area now covered by woodland, the UK has become one of the least forested nations in Europe. Worse still, 97% of the UK's wildflower meadows, a rich source of nectar for insect pollinators, has been lost since the 1930s, along with 200,000 miles of hedgerows with flower-rich field margins (enough to go eight times around world). And it is estimated that 80% of our calcareous grassland, home to many rare and interesting plants, has been lost over the past 60 years. Most of the good habitat that remains is in small, isolated fragments and continues to be vulnerable. Losses of flowers in all habitats is often exacerbated by poor management. Grasslands and heathlands are being overgrazed, or sometimes undergrazed (which can be just as bad), and the excessive numbers of deer that roam woodland is causing serious damage to its understorey and its flowers.

Wildflower meadow, County Durham [PC]

Monoculture being harvested [pxhere.com]

The decline of insect pollinators in the UK that has occurred in the decades since the end of the Second World War is undoubtedly a result of a lack of flowers caused by modern farming—notably the reseeding of previously species-rich pastures, the removal of hedgerows and woodland to make bigger fields, intensive grazing and the regular application of inorganic fertilisers and herbicides. So-called 'improved pastures' that used to be semi-natural grassland, full of wildflowers, are now so 'improved' that they are little more than green deserts, almost devoid of flowers and nectar for pollinators.

A particularly important reason for the scarcity of insect pollinators in farmland is because modern intensive agriculture creates monocultures that, combined with copious applications of herbicides, eliminate the insect-friendly flowers that used to mingle with crops. Many monocultures, including wheat, corn and other 'grasses' have wind-pollinated flowers and provide nothing for Honey Bees, wild bees or other pollinators. Even though other monocultures, such as Oil-seed Rape, sunflowers and orchards, do provide nectar and pollen, at best they flower for just a few weeks, ensuring that there is little or nothing for pollinators during the rest of the year. Because pollinators are lacking, insect-pollinated monocultures often have to rely for pollination on hives of Honey Bees that have to be moved around the countryside as needed. This has been shown to have adverse effects on the bees' health—being moved frequently causes stress; bees forced to feed just on pollen from a monoculture suffered nutrient deficiencies and have a less healthy immune system than bees feeding on pollen from a variety of plant species; and stressed bees and their larvae are less able to withstand microbes and other pathogens.

These major changes in British farming systems have resulted in a devastating dearth of nectar and pollen for bees and other pollinators. A study, by Laura Jones and colleagues compared Honey Bee foraging behaviour in 1952 with 2020. Using DNA barcoding to analyse honey samples, the study revealed significant changes in the wildflowers available to Honey Bees (and other pollinators). In the 1950s, flowers were still abundant in pastures and white clover was the single most important source of pollen and nectar. But, nowadays, semi-natural grassland is estimated to be only 3% of what was present prior to the Second World War and farmland is now so lacking in flowers that Honey Bees have been forced to forage elsewhere to seek out inferior alternatives, notably Bramble, Oil-seed Rape (that flowers for only a few weeks each year) and Himalayan Balsam *Impatiens glandulifera* (an invasive alien and noxious weed of waterside habitats). 'Improved' grasslands are a dominant habitat in the UK and changes to the way they are managed have the greatest potential to enhance the floral resources in the British countryside. The restoration of flower-filled grasslands and hedgerows would massively improve the supply of nectar and pollen for bees and other pollinators.

Losses of pollinators and other insects are not confined to farmland; the losses are also occurring in nature reserves, national parks and other areas of apparently excellent habitat. The results of a German study, for example, carried out by the Krefeld Entomological Society, are worrying because the study was carried out in nature reserves, not on farmland. Between 1989 and 2014, the Society monitored flying insects—bees, flies, beetles, butterflies and moths—in over 100 nature reserves in western Europe and found that both insect biomass and diversity had plummeted by almost 80%. In 1989 they collected 17,291 hoverflies of 143 species. In 2014, in the same reserves, they collected only 2,737 specimens of 104 species.

These findings come as no great surprise to those of us old enough to remember British insect life in the 1950s and 60s, in the days before farming destroyed so much natural habitat. In his beautiful book, *The Moth Snowstorm: Nature and Joy*, Michael McCarthy remembers summer nights in his boyhood when moths *"would pack a car's headlight beams like snowflakes in a blizzard"*. Many of us still have the same memories—the way moths and other insects used to smear car windscreens and headlights, requiring regular stops to wipe them clean. I also remember the woolly bears—the caterpillars of the Garden Tiger moth *Arctia caja*—and other caterpillars that I encountered as they wandered in search of sites to pupate. As a young child I was fascinated by caterpillars—the more colourful and ornamented the better. I reared many and even dreamed about them. Nowadays, I almost never encounter the woolly bears and other fancy caterpillars that used to be so common and I recently learned that the once common Garden Tiger has declined by 92% since the 1960s. And, of course, huge losses of insects have knock-on effects up the food chain. It is hardly surprising that many insect-eating birds are declining.

The moths and 'car windscreen phenomenon' has been noted anecdotally by numerous entomologists. In 2004, this prompted the RSPB to investigate the phenomenon by asking motorists (40,000 of them nationwide) to use their front number plate as a 'splatometer' and count the splattered corpses. It was a surprise that only 324,814 splats were recorded—an average of only one splat for every five miles. The study was prompted by fears that decreasing insect populations must cause problems for those bird species which rely on insects for food. In fact, the recent 2019 RSPB *State of Nature* report, which brings together findings from 50 organisations, suggests there has been a shocking 59 per cent decline in insects in the UK since 1970.

Garden Tiger *Arctia caja* caterpillar ('woolly bear') and moth

CHAPTER 5
Intensive agriculture and chemicals

Other than habitat loss, the most important reason why pollinators are declining in the UK and elsewhere is probably the excessive use of insecticides, herbicides and other agricultural chemicals. The ability of bees of all species to withstand parasites and diseases is strongly influenced by the quality of the food they consume and how badly it is contaminated. Nowadays, bees foraging on farmland are exposed to a toxic 'pesticide cocktail' in the pollen that is the most important source of nutrition for both their colonies and broods. In one important study scientists identified traces of over 150 harmful chemicals in the pollen collected by bees, and consistently found several different chemicals in almost every sample. The survey, which involved 23 states in the USA and a province in Canada, concluded that *"surviving on pollen with an average of seven different pesticides seems likely to have consequences"*—an understatement if ever there was one. In fact, the quantity and number of chemicals that are applied to crops is hard to comprehend. Felicity Lawrence, a British journalist, and author of two important exposés concerning the food business—*"Not on the Label"* and *"Eat Your Heart Out"*,—wrote in *The Guardian* (Monday 3 October 2016) *"British farmers growing wheat typically treat each crop over its growing cycle with four fungicides, three herbicides, one insecticide and one chemical to control molluscs. They buy seed that has been precoated with chemicals against insects. They spray the land with weedkiller before planting, and again after. They apply chemical growth regulators that change the balance of plant hormones to control the height and strength of the grain's stem. They spray against aphids and mildew. And then they spray again just before harvesting with the herbicide glyphosate to desiccate the crop, which saves them the energy costs of mechanical drying"*. Similar regimes of chemical spraying are found throughout most of the developed world. It is also worth mentioning that the herbicide glyphosate is now so commonly used in Europe that it has been detected in the urine of 44% of people surveyed for Friends of the Earth.

In recent years, neonicotinoid insecticides (often shortened to 'neonics'), have received much publicity in the media. Neonics are neurotoxins chemically similar to nicotine and extremely potent. It has been reported that four-billionths of a gram is a lethal dose for a bee which translates to one teaspoon of neonics being sufficient to kill one and quarter billion bees. Traces of most traditional insecticides linger on the foliage of plants for a long time but neonics work differently. Neonics are systemic insecticides, meaning that they are taken up by plants, ensuring that any animal that swallows bits of any treated plant also swallows the insecticide. It is also noteworthy that no more than about 5% of the insecticide is taken up by the plant. The remaining 95% eventually accumulates in soil, groundwater, or is absorbed by other plants, in all of which places it is an ongoing threat to the environment and its wildlife. One worries especially about the effect of neonics on the microscopic soil fauna that plays a critical role in maintaining soil health and fertility. Bear in mind that a single handful of healthy soil contains literally billions of organisms—bacteria, fungi, algae and protozoa, as well as earthworms and other larger animals—many of which contribute to soil health.

Many agricultural chemicals have harmful consequences for bees and other pollinators even if they are not killed immediately. For example, low doses of neonics are thought to badly affect bees' resistance to disease as well as their ability to navigate and recognise flowers. A recent study found traces of neonics in 75% of honey samples from across the world. Scientists say the levels are far below the maximum permitted levels in food for humans. Even so, in one-third of the honey samples, the amount of the chemical found was enough to be detrimental to the bees that would have eaten the honey. And yet another study has shown that neonics and organophosphates disorient and sicken migrating songbirds. And are, therefore, likely to be detrimental to other vertebrates.

The results of some very recent research, carried out by scientists at Imperial College, London and published in August 2018, are even more interesting (and shocking) and add to the evidence that neonics harm wild bees. According to scientist Dr Andres Arce, "*Many studies on neonicotinoids feed bees exclusively with pesticide-laden food, but in reality, wild bees have a choice of where to feed. We wanted to know if the bees could detect the pesticides and eventually learn to avoid them by feeding on the uncontaminated food we were offering. Whilst at first it appeared that the bees did avoid the food containing the pesticide, we found that over time the bumblebees increased their visits to pesticide-laden food.*" In other words, bumblebees seemed to become addicted to contaminated food containing neonics, just as humans become addicted to the nicotine in cigarettes.

Neonicotinoids continue to be a major problem. There have been recent news reports (August 2019) about mass die-offs of Honey Bees in numerous countries, including Brazil, the USA, Canada, Mexico, Argentina, Russia, Turkey and South Africa. In Brazil, for example, more than 500 million Honey Bees died within three months in 2019. It is thought the die-offs were caused by the use of pesticides, including neonicotinoids and fipronil (a broad-spectrum insecticide that disrupts the insect central nervous system), that are banned by the European Commission (the executive branch of the European Union). Fipronil was also blamed for die-offs in South Africa. In the USA, the last year (2018–19) has been the worst on record for Honey Bee die-offs with the loss of 40% of bee colonies.

In 2018, several neonicotinoids were banned by the EU and UK because of the serious damage they could do to bees. At the time of the ban, Michael Gove, then environment secretary, said the UK was in favour as it could not *afford to put our pollinator populations at risk*. Unfortunately, now in 2021, one of the banned neonics—thiamethoxam—has been authorised for emergency use in England (and elsewhere in the EU), despite the earlier ban. Its use has been

Honey Bee visiting Gorse *Ulex europaeus* [PC]

allowed because a virus—beet yellows virus which is transmitted mainly by aphids—seriously reduced the amount of sugar beet grown in 2020. However, the Wildlife Trusts are challenging this decision and taking legal action against the UK government for reversing its earlier ban. Professor Dave Goulson, a noted bee expert at the University of Sussex, has also said that it makes no sense to support the sugar industry because we "*consume far too much, and it is a significant contributor to the obesity epidemic and associated surge in diabetes.*"

Some of us are old enough to remember the publication, in 1962, of Rachel Carson's book *Silent Spring*. Her book described the harmful effects of pesticides on the environment in the USA and led, in 1972, to a nationwide ban on the use of DDT. Between 1968 and 1980, the use of DDT was banned in several other, mainly European, countries (though not in the UK until 1984). Even so, DDT was still being manufactured and used, or sold elsewhere, in China and India well into the 21st century. *Silent Spring* also highlighted how human activity so often adversely affects the natural world. Nevertheless, today, almost 60 years later, the lessons of *Silent Spring* have still not been learned—nothing much has changed. The natural world is still being ravaged by agricultural chemicals—fertilisers, herbicides and pesticides, including neonicotinoids, as well as by continuing habitat destruction, pollution of rivers and oceans, and climate change.

It is also worth mentioning at this point that, even now, the UK is a major exporter of pesticides that are banned in the EU. Critics in the importing nations claim, with some justification, that the practice is a 'double standard' that places a lower value on lives and ecosystems in poorer countries. According to the environmental group Greenpeace, using freedom of information requests to obtain the data, the UK and other European countries exported 81,615 tonnes of banned chemicals in 2018, of which the UK was responsible for more than 32,000 tonnes (40%). Greenpeace described the export of chemicals banned in the UK as "*exploitative hypocrisy*". Most of the UK's exports contain paraquat, a weedkiller banned in the EU since 2007. It is an effective weedkiller but can be extremely toxic in concentrated doses, so much so that it is often used in suicides. Repeated exposure has also been linked with Parkinson's disease.

CHAPTER 6
Hi-tech indoor and rooftop farming

Thanks to hydroponics, artificial lighting and heating, such crops as tomatoes, lettuce, cucumbers, peppers, spinach, strawberries and blueberries are now often grown indoors in city centres and other urban areas. An indoor growing environment avoids many of the problems suffered by conventional farming. It provides perfect, 365 days a year, growing conditions, without any danger of the unpredictable weather events—heat waves, severe freezes, torrential rain or droughts—that are becoming increasingly common due to climate change (e.g. 2018's 'beast from the east' and 2019's February heat wave). The growing environment is also pesticide-free, uses 70% less water than traditional farming systems and, given its inner city location, drastically reduces food miles for retailers and consumers.

For example, the British company Growing Underground is said to be the world's first underground urban farm. The company, which is based in London, makes use of Second World War bomb shelter tunnels, 33 metres under the busy streets of Clapham. Using the latest hydroponic systems and LED technology, it grows salad leaves, micro greens and herbs that are supplied to many of London's supermarkets and restaurants. One would hope that similar spaces in other cities, such as underground car parks, abandoned warehouses and disused tunnels, will be utilised in the same way. In fact, indoor farming is already expanding rapidly in the USA, Japan, China and elsewhere, mainly in the form of vertical farming—farming in tall, city buildings, using all the building's floors, not just the rooftop. In vertical farming, one hectare of indoor space is the equivalent of five or more outdoor hectares, so less space is needed to grow an equivalent amount of produce. Indoor vertical farming can be very high-tech—fully automated with a computerised operating system that controls and monitors just about everything, including seeding machines, lighting, nutrient flow, plant health and crop harvesting. It has been estimated that the global value of the vertical farming market will grow from $403 million in 2013 to $6.4 billion by 2023.

Indoor vertical farming in city centres or other urban areas is particularly suited to perishable produce that has a short shelf life and doesn't travel well. Crops such as tomatoes and strawberries could even be bred and grown to enhance their taste rather than their ability to survive long travel times. On the other hand, indoor farming does not make much sense for crops with a long shelf life, most of which can be grown more efficiently and cheaply outdoors on traditional farms. Indoor vertical farming takes place in a controlled, enclosed environment which leads to one major problem—no natural pollinators for crops that need to be cross-pollinated. Although automation is the norm for many of the crops grown in vertical farms, which requires fewer workers, some crops may need to be pollinated manually, which is labour intensive and hence expensive. However,

Growing food underground [independent.co.uk]

hives of bumblebees are now raised and supplied commercially for use as pollinators in greenhouses or other enclosed spaces. Usually, their diet needs to be supplemented with sugar water or 'bee candy'. And it is important not to use pesticides on the plants to be pollinated.

An even more recent hi-tech development is the proposal to help minimise carbon emissions by using waste heat, generated by wastewater processing in sewage farms, to heat greenhouses in which tomatoes, cucumbers, peppers and other produce can be grown all year round. The company involved—Low Carbon Farming—is currently building two huge greenhouses near Norwich and Bury St Edmunds, covering 16 hectares and 13 hectares respectively, that will operate using heat from sewage processing systems operated by Anglian Water. The greenhouses will be the first in the world to take waste heat from sewage farms and, using ground source heat pumps, pump the surplus heat to the greenhouses. Over the next five years, Low Carbon Farming plans to build 43 sewage heated greenhouses at an estimated cost of £2.67 billion. The greenhouses will operate using hi-tech methods, including hydroponics and artificial intelligence. Presumably, bumblebees will be used to take care of pollination, as they are in some other commercial greenhouses (see above). There will be many benefits. It is estimated that the carbon footprint will be reduced by 75% compared with the usual gas-heated greenhouses; screens will drastically cut the need for pesticides; the recirculation and conservation of water will mean that 10 times less water is needed than in conventional farming in open fields; and the crop yield will be 10 times more than is grown in an equal area of fields. The first tomatoes are expected to reach supermarkets by spring 2021. Once fully operational, the 43 greenhouses should produce 600,000 tons of tomatoes annually, more or less equal to annual demand in the UK. Currently, 80% of tomatoes have to be imported and only 20% are home-grown. Given the new greenhouses, imports of other crops, such as peppers, cucumbers and soft fruits, should also be unnecessary or at least greatly reduced.

Greenhouse tomatoes [f8studio/123rf]

Rooftop farm, Hong Kong [gogreenhongkong.com]

Urban rooftop farming is also in the news. Rooftop gardens have been around for years but rooftop farms are relatively new. A company in Paris—Agripolis—is currently (in 2020) renovating the Expo Porte de Versailles creating what will soon become the world's largest urban rooftop farm, with an area of 14,000 square metres, equivalent to about two soccer pitches. About 30 different crops will be grown in vertical stacked columns without soil and fed with nutrient-rich solutions and rainwater. The farm, which will be organic and pesticide and chemical free, is expected to produce around 1,000 kg of fruit and vegetables every day when growing conditions are favourable. Founder, Pascal Hardy, envisions *"a city in which flat roofs and abandoned surfaces are covered with these new growing systems. Each will contribute directly to feeding urban residents who today represent the bulk of the world's population"*.

CHAPTER 7
Global warming

The world is getting warmer. The burning of fossil fuels, such as coal, oil and gas, in ever-increasing amounts over the last two centuries, has released enough carbon into the atmosphere to cause a 'greenhouse effect' and the global warming that has followed has resulted in unpredictable 'climate chaos'—rising sea temperatures and sea levels; melting ice in the Arctic and Antarctic; melting permafrost; hotter, drier summers and droughts; wetter winters and flooding; stronger trade winds and catastrophic storms and hurricanes. As of 2020, the 20 warmest years on record have been in the past 22 years and, according to global meteorological services, 2020 was either the hottest, or tied with 2016 as the hottest year on record. If the warming trend continues, temperatures may well rise by 3–5°C by 2100. According to the Intergovernmental Panel on Climate Change (the IPCC), the world will face calamitous consequences for biodiversity, agriculture and food production if mankind fails to reverse this trend. The IPCC recommends that the rise in average global temperature should be limited to no more than 1.5°C above pre-industrial levels. Global warming is a particularly serious threat to the phenology of plants and how their flowering matches the seasonal distribution and abundance of pollinators. Global warming is likely to induce mismatches in the spatial and temporal distribution of flowering plants and their pollinators.

At the present time, in November 2021, the outlook continues to be grim. Extreme weather in 2020 saw thousands of lives lost and enormous insurance costs. For example, hurricanes and wildfires in the US caused some $60 billion in losses, while flooding in India resulted in more than 2,000 deaths and insurance losses of $10 billion. Super-cyclone Amphan slammed into the coast of eastern India and Bangladesh in May, bringing gales and flooding, causing over 100 deaths and estimated losses of $13 billion in just a few days. China suffered even greater financial damage from flooding—around $32 billion between June and October. Africa also suffered, with heavy rains resulting in massive locust swarms that ruined crops to the value of $8.5 billion. Since 1980, when records began, climate disasters have been becoming both more frequent and more costly. According to the US National Oceanic and Atmospheric Administration, these disasters have cost the US economy $1.875 trillion in the last four decades.

Less catastrophic weather events in 2020 were common elsewhere in the world. Temperatures in the Siberian Arctic were more than 5°C above average between January and June 2020 and the region is believed to be getting warmer at least twice as fast as the rest of the world. And, in June, the region even recorded an extraordinary temperature of 38°C, the highest temperature ever recorded north of the Arctic Circle. An international team of climate scientists, led by the UK Met Office, concluded that the unprecedented average temperatures in the Siberian Arctic were *"unequivocal evidence of the impact of climate change on the planet"* and likely to occur less than once in 80,000 years, without the influence of human induced climate change—warming resulting from worldwide greenhouse gas emissions (e.g. carbon dioxide, methane and nitrous oxide, both the latter two gases being many times more potent than carbon dioxide).

This conclusion is significant because *"what happens in the Arctic does not stay in the Arctic"*. The Arctic is one of the major drivers of global weather and is likely to cause significant ripple effects around the world, leading to extreme weather events. Notable weather events in the UK in the last few years include the 'beast from the east' in February-March 2018 that caused over a billion pounds worth of damage. And the year 2020 saw disastrous floods in February; a record breaking dry, sunny spring and early summer; followed by the wettest October in well over 100 years.

The years 2019 and 2020 saw numerous dramatic weather events. One of the most exceptional was the unprecedented melting in 2019 of nearly 600 billion tonnes (a tonne is a metric ton = 1,000 kg) of Greenland's massive ice sheet, more than double the average losses between 2002 and 2019. The 2019 losses are equivalent to about a million tonnes a minute and the biggest single contributor to current sea level rise. And, in 2020, a huge chunk of ice from the Arctic's largest remaining ice shelf, measuring 110 sq km in north-east Greenland, broke away and disintegrated into tiny pieces. Polar scientists say that this is yet more evidence of the rapid climate change occurring around Greenland. The region has warmed up by about 3°C since 1980 and experienced record summer temperatures in 2019 and 2020. Recently (in October 2020), the German Research Vessel Polarstern arrived back in its home port after spending a year in the Arctic Ocean studying the changing Arctic climate. Expedition leader, Professor Markus Rex, returned with a warning. "*The ice is disappearing and if in a few decades we have an ice-free Arctic - this will have a major impact on the climate around the world.*"

Another disturbing development in the Arctic has been documented by an international team of scientists aboard the Russian research ship R/V Akademik Kelysh. The researchers found that frozen methane deposits in the Arctic Ocean are being released in shallow seas over the East Siberian Arctic Shelf. Some of the released gas dissolves in the sea but methane concentrations at the surface are four to eight times higher than normal and some methane is venting into the atmosphere. Methane is a much more potent greenhouse gas than carbon dioxide. Over a 20-year period, it traps 84 times more heat than the same mass of carbon dioxide.

Also, between January and the end of August 2020, wildfires in Siberia were estimated to have released 244 million tonnes of carbon dioxide, a figure a third higher than 2019's previously record levels which in turn dwarfed totals from earlier years. The situation throughout Siberia is aggravated by so-called 'zombie fires' that smoulder throughout the winter in peat below the frozen surface of the ground, and then reignite in spring, burning not only the forest but also the underlying peat with its gigantic store of carbon dioxide. The record temperatures in Siberia have also caused early melting of sea ice and widespread thawing of the permafrost. The thawing permafrost is set to have other devastating consequences. Researchers now believe that for every one degree Celsius rise in the global average temperature, thawing permafrost may release the equivalent of four to six years' worth of coal, oil, and natural gas emissions, which is double or triple what scientists were expecting a few years ago. And if our use of fossil fuels is not drastically reduced, thawing permafrost could be as big a source of greenhouse gases as China is now. Globally, permafrost holds up to 1,600 billion tonnes of carbon, nearly twice as much as is currently in the global atmosphere. Much of the carbon is in the form of methane—a greenhouse gas that is many times more potent than carbon dioxide. It is a scary statistic.

A recent report in Nature Climate Change (2021) is perhaps even more alarming. Between 1955 and 1990, the Soviet Union scuttled over 100 decommissioned nuclear submarines in the vicinity of the Novaya Zemlya archipelago (in the Arctic Ocean to the north-east of Scandinavia) and also carried out nuclear weapons tests that released about 265 megatons of nuclear energy. Following these actions, it is now thought that warming Arctic seas and melting ice might well lead to the release and spread of nuclear waste and radiation. Other potential pollutants that may be released by melting permafrost include mining by-products, such as arsenic, mercury and nickel, and antibiotic resistant microorganisms. Melting ice and increased water circulation risks pollutants, some of them radioactive, entering the human food chain as well as damaging wildlife. The Russian government has launched a clean-up plan.

Thawing ice in the Antarctic is also set to cause problems by being a major contributor to the rise of global sea-levels. For example, the Thwaites Glacier in western Antarctica (a glacier almost as big as Britain) is shedding icebergs and melting at an ever increasing rate. In the 1990s it was losing about 10 billion tonnes of ice a year, a total that is now closer to 80 billion tonnes. Evidence found by researchers from the UK and USA suggests that deep seafloor channels are carrying warmer water from the depths of the surrounding ocean to the base of the melting

Thwaites Glacier, Antarctica [Picasa (pxherre.com)]

glacier. In time, the Thwaites Glacier alone has the potential to add 65 cm to global sea-levels—enough to inundate some of the world's low-lying cities.

Also notable in 2020, was the severe heatwave that resulted in a blistering, world record high temperature of 54.4°C (130°F) in Death Valley in California. Exacerbated by the heatwave and strong, hot, dry winds, wildfires in California, Oregon and Washington burned about 2.7 million hectares (27,000 sq km) of forest and scrub by mid-September 2020. The statistics are disturbing. The 900 wildfires still burning in mid-September 2020 covered an area six times bigger than the total of all of 2019's fires added together. And, in California alone, more than 17,000 firefighters were involved in combating the unprecedented fires—unprecedented yes, but long forecast by climate scientists.

Human-induced climate change has already led to a worldwide increase in the frequency and severity of fire weather, increasing the risk of devastating wildfires, such as those that occurred in mid-2020, in the western US, southern Europe, Siberia, Australia and Amazonia. As far as the western US is concerned, President Donald Trump blamed the wildfires on poor forest management by the states involved (i.e. California, Oregon and Washington), not on climate change, and added "*I don't think science knows*" what is actually happening. It is somewhat ironic, therefore, that it is the federal government, not the individual states, that owns most of the forest that has gone up in flames. In California, for example, the federal government (including the US Forest Service, the Bureau of Land Management and the National Parks Service) owns nearly 58% of the 33 million acres of forested land in the state and is responsible for its management. California itself owns no more than 3% with the rest in private hands. Ownership and management of forest in Oregon and Washington is also mainly in federal hands. It is also relevant that there have been extensive funding cuts to relevant federal agencies under President Trump,

Scientists are agreed that poor forest management alone is not responsible for the recent infernos. A team of scientists from the UK and Australia have reviewed (in "*Climate Change Increases the Risk of Wildfires*") more than 100 studies, published since 2013, that show that extreme fires occur when natural variability in the climate is superimposed on the increasingly warm, dry and windy conditions that result from global warming. The same research team reviewed the reasons behind Australia's dreadful fires in the 2019–2020 seasons and reached a similar conclusion. The researchers acknowledge that fire management practices in the US have contributed to the build-up of fuel. Normally, controlled burns are carried out to reduce the amount of fuel available when a wildfire strikes. But prescribed burns can only be carried out when conditions are not too hot, dry and windy, because only then is it possible to control the fires. The researchers concluded that conditions favouring wildfires are likely to continue into the future and the resulting fires may well get worse.

Elsewhere in 2020, the rate of destruction of the Amazon rainforest in Brazil has reached its highest level since 2008. More than 11,000 sq km of rainforest were destroyed, by burning or logging, in the year from August 2019 to July 2020—a 9.5% increase over the previous year. According to the Brazilian Institute of Space Research (INPE), there were 17,326 fires blazing in the Amazon in October 2020, compared to 7,855 in October 2019. Satellite data also suggests that there were a record number of fires in the Pantanal wetlands at the same time. Scientists and environmentalists have accused the Brazilian president, Jair Bolsonaro, of deliberate deforestation—encouraging the destruction of rainforest to facilitate mining activities and to create farmland for cattle or crops, particularly soya. President Bolsonaro has also cut funding to the federal agencies that have the power to fine and arrest miners, loggers and farmers for breaking environmental laws. The Brazilian Space Research Institute revealed data that documented the increasing deforestation in the Amazon—an increase described as a "*lie*" by President Bolsonaro, who fired the Institute's Director. The Brazilian office of Greenpeace has also accused the current Brazilian administration of trashing almost all the work done in recent decades to end deforestation in the Amazon. The lack of protection for the Amazon rainforest and its indigenous inhabitants and their traditional settlements is a tragedy. The Amazon region's billions of trees are a vast carbon store and a vital defence against global warming that, in a typical year, absorbs about 1.5 gigatons of carbon dioxide from the atmosphere. However, a recent study (Qin *et al.*, 2021) showed that, as a result of climate change and deforestation, the Brazilian rainforest released about 20% more CO_2 into the atmosphere than it absorbed during the years 2010-2019. There are calls for President Bolsonaro to face charges in the UN-backed international criminal court for 'ecocide' and crimes against humanity.

Amazon rainforest destruction, Brazil

2021 wildfires, Oregon [Oregon State Fire Marshall/nytimes.com]

Among many weather related catastrophes in 2021, one of the most extraordinary was the heatwave experienced in late June in the Pacific north-west of the USA and adjoining Canada. The region, which is known for its normally mild summer climate, experienced an unprecedented 'heat dome' that smashed temperature records, often by 4–5°C. It is estimated that over a billion marine creatures were killed along the Pacific coast—oysters, mussels, clams and sand dollars—all 'cooked' by the exceptional temperatures. The small Canadian town of Lytton set an astonishing new Canadian record temperature of 49.6°C (121.3°F) and the next day was almost totally destroyed by a fast-moving wildfire. Climate scientists have said that the heatwave would have been *"virtually impossible"* but for human-induced climate change. The North American wildfire season in 2021 was predicted to be particularly severe. As of September 2021, the National Interagency Fire Centre (NIFC) reported that 2.7 million hectares of land had been burned by over 44,000 wildfires in the USA, the largest of them in California and Oregon. Particularly newsworthy was the fire in Sequoia National Park at the end of September 2021 that threatened to burn the 83 m tall General Sherman tree, the biggest tree by volume on Earth and about 2,500 years old. Firefighters protected the tree, and many others, by wrapping the tree's base with fire-resistant aluminium foil. This summer's exceptional heat has not been restricted to North America. In Russia, a heatwave matched a 120-year-old record temperature; Northern Ireland broke its temperature record three times in the same week; and a new high was set in Antarctica.

Extreme weather events have continued throughout much of 2021. The earth's warming atmosphere holds more and more moisture, often resulting in the torrential rainfall and severe flooding that have been seen in many parts of the world, including Western Europe, India, China, Japan, New Zealand and elsewhere. Parts of Western Europe, particularly in Germany, were deluged with up to two months worth of rain that fell in only two July days. More than 100 people died and property damage was devastating. Madagascar is experiencing the opposite catastrophe. According to the United Nations, southern Madagascar is on the brink of experiencing the world's first *"climate change famine"* brought about by a four-year drought—the worst in four decades. Tens of thousands of people, who don't burn fossil fuels and have not contributed to climate change, are on the brink of famine.

Global warming does not mean that everywhere in the world gets warmer at a uniform rate. But it does amplify the risk of extreme weather disasters. All the extra energy in the atmosphere destabilises weather systems, resulting in extreme, freakish weather events that are both more frequent and more unpredictable than they used to be and often localised. As indicated above, the scale of destruction all over the world is both new and horrifying. These extreme events include deadly heat waves and wildfires; torrential rainfall and record flooding; intense prolonged draughts (e.g. Chile's mega-drought which has now continued for 10 years) resulting in repeated crop failures; and periods of intense cold in normally temperate areas, occasionally with temperatures as low as those in the Arctic or Antarctic. One consequence of all the extra energy in the atmosphere is that North Atlantic hurricanes are now even more damaging and dangerous after making landfall. The time it takes hurricanes to weaken over land has almost doubled in the past 50 years. Hurricane damage is expected to get even more severe in the future.

The frequency of extreme weather events has increased five-fold in the past 50 years and the United Nations Environment Program estimates that coping with the damage will cost developing countries US$140-300 billion per year by 2030. As far as the UK is concerned, in 2020 it experienced its wettest ever February, a record sunny May and the wettest-ever day on 3 October. To cope with these now expected climate emergencies, the UK Met Office is launching a climate analysis tool to help planners prepare for future extremes of rainfall and high temperatures. It warns that wild weather is likely to place increasing challenges on health, infrastructure and services.

There is now little doubt that the changing climate is a factor that is contributing to pollinator declines. In the UK, many plants now flower 2–3 weeks earlier than was the norm 20 years ago. As climatic conditions continue to change, and unpredictable weather becomes more common, there is likely to be a crucial lack of synchronisation between the timing of flowering and the emergence of pollinators in spring. Plants that depend on a single pollinator, such as Yellow Loosestrife *Lysimachia vulgaris*, White Bryony *Bryonia dioica* and Ivy *Hedera helix* (pollinated by Yellow Loosestrife Bee *Macropis europaea*, Bryony Mining Bee *Andrena florea* and Ivy Bee *Colletes hederae* respectively), and those with a short, early flowering season, including willows *Salix* spp., Blackthorn *Prunus spinosa*, apples *Malus* spp. and other fruit trees, are likely to be vulnerable to a lack of suitable pollinators on the wing at the critical time.

Recent research shows that the unpredictable freakish weather that now prevails—climate chaos—has caused widespread losses of bumblebees (among the most important pollinators) across Europe and North America. The researchers looked at more than half a million records of 66 bumblebee species and found that, since the 1970s, the likelihood of any locality being occupied by bumblebees had declined by over 30%. They found that the declines are best explained by the increasing frequency of extremely hot days rather than the relatively small increase in average temperatures. It is alarming, therefore, that heat waves are expected to continue to become both more frequent and intense as a result of global warming. For the UK, a heat wave is defined as a period of at least three consecutive days with maximum temperatures meeting or exceeding a temperature threshold in the range 25–28°C (with different thresholds for different parts of the country). In fact, bumblebees have already been hit particularly hard in countries such as Spain and Mexico that are now subjected more frequently to exceptionally hot years. And two recent summers (2018 and 2019) saw exceptional heat in the UK and Europe. A new UK record temperature of 38.7°C was set in Cambridge University Botanic Garden in July 2019 and it is now thought that temperatures of up to or over 40°C could be recorded regularly in the UK by 2100. Temperatures of over 40°C have already been recorded in countries as close to the UK as Holland and Belgium.

Climate change 'deniers' who claim that global warming is a result of natural cycles have been proved wrong by a new study published in the prestigious journal Nature. According to the study, global warming is happening at an "*unprecedented*" rate that far exceeds any temperature fluctuations seen during the last two millennia. The warming recorded in the 20th century, in over 98% of the world, is in "*stark contrast*" to previous extreme cold or warm episodes (e.g. Britain's 'Little Ice Age' that occurred from the early 14th century through the mid-19th century). The Little Ice Age occurred on a local regional scale, not synchronously on a worldwide scale.

In another recent paper published in Nature, climate scientists discussed the potential tipping points that are threatening the world's climate. Climatic 'tipping points' are points at which even a small climatic change can lead to an abrupt, runaway change that is more or less irreversible. Examples of potential tipping points include the runaway melting of the Greenland and Antarctic ice sheets, which would eventually lead to a sea-level rise of over 60 m; the destruction of the Amazon rainforest which would add to greenhouse gas emissions and their impact on global warming; and the collapse of the Atlantic conveyor, leading to serious changes in climate and rainfall patterns in the northern hemisphere.

Despite all the dire warnings from climatologists and the United Nations, the world remains on course to suffer the consequences of intensifying climate change. As a result of mankind's use of fossil fuels (coal, oil and gas), the past decade was the warmest on record, resulting in extreme weather events. With a rise of just 1.1°C, the world has endured prolonged heatwaves; droughts and crop losses; wildfires fuelled by desiccated vegetation; and torrential rainfall and flooding. Scientists have recently said that global carbon emissions ought to be reduced by 45% by 2030. Instead, it is currently expected that emissions will actually go up by 16% over the same period—an increase that is a major cause for concern, especially for the world's most vulnerable people. Governments agree that urgent collective action is needed.

The end of 2021 has brought COP26 (the 26th United Nations Climate Change conference held in November in the city of Glasgow, Scotland). The Glasgow meeting has been seen as a crucial opportunity to curb ongoing climate change. Some progress has been made. However, the Climate Action Tracker (CAT) calculates that the world is heading towards 2.4°C of warming rather than the 1.5°C limit considered crucial. It also considers that COP26 "*has a massive credibility, action and commitment gap*" and that the world is not close to limiting the global temperature rise to an acceptable 1.5°C.

The goal of the numerous COP meetings, past and present, has been to keep cutting emissions until they reach net zero, hopefully by 2050. Progress continues but the critical question is— what will the climate be like when we finally reach net zero? To keep warming limited to 1.5°C by 2100, scientists say that carbon emissions will have to be slashed dramatically by 2030. COP26 has made recommendations in five important categories: US-China co-operation, coal, methane, trees and money. **1.** The most unexpected development was probably the announcement that **THE US AND CHINA** (the world's two biggest carbon dioxide emitters) have decided to work together to achieve the 1.5°C temperature goal set out in the 2015 Paris Agreement. To this end both countries have agreed on various issues, notably the transition to clean energy. **2.** With regard to **COAL**—the single biggest contributor to climate change which was still producing about 37% of the world's electricity in 2019—more than 40 countries committed to shift away from coal, though several major producers and/or users, including China, India, Australia and the US, did not sign up. **3.** In the case of **METHANE**, a potent greenhouse gas responsible for a third of man-made warming, more than 100 countries agreed to cut 30% of their current methane emissions by 2030. Again, China, Russia and India—the biggest methane emitters—have not yet signed up. **4.** Some progress has been made on **TREES AND DEFORESTATION**. A total of 110 countries (that control about 85% of the world's forests) promised to stop deforestation by 2030. This is important because the UN says 420 million hectares of forest have been lost since 1990 and trees help to tackle climate change by absorbing enormous amounts of carbon dioxide—a key greenhouse gas. However, it is not clear how the promises will be policed, particularly after Indonesia described the proposal as "*unfair*" and a deal struck in 2014 failed to slow deforestation at all. **5.** Finally, there is the question of **MONEY**. As many as 450 financial organisations—banks, insurers and pension funds—that control assets worth $130 trillion, have agreed to back renewable energy rather than technology that burns fossil fuels. This initiative is intended to provide finance for green technology and help meet net zero targets. Unfortunately, in the past, the trust of poorer nations was damaged when richer nations failed to provide the $100 billion that had been promised by 2020.

PART 3
Saving pollinators

Shrill Carder Bumblebee *Bombus sylvarum* [PB]

CHAPTER 8

Reversing pollinator declines in farmland

So, what can be done to make matters better for pollinators, particularly in farmland? In addition to limiting the effects of global warming, prohibiting the most harmful pesticides, herbicides and other agricultural chemicals would be a good start. It would also help to follow the recommendations of the Rio Declaration (sponsored by the United Nations Environment Program and signed by 172 countries in 1992) and its 'precautionary principle', defined as follows: "*Where there are threats of serious or irreversible damage, lack of full scientific certainty shall not be used as a reason for postponing cost-effective measures to prevent environmental degradation.*" Regrettably, governments often disregard the precautionary principle, fearing that banning agricultural chemicals might have harmful consequences for both farmers and consumers. It is a pity that democratic governments, elected for just a few years, too often opt for short-term, political gains, rather than long-lasting benefits.

Given prevailing attitudes, it is gratifying that in April 2018 European Union Member States voted unanimously for a near-complete ban on neonicotinoid insecticides. The UK, along with the Netherlands, Ireland, Germany and Austria, had previously resisted any major extensions to the restrictions in force since 2013. Unfortunately, it will still be possible to use neonics in greenhouses, in spite of the danger of the chemicals leaching into the water table.

In another promising development, 'sexy plants' are well on the way to replacing many pesticides. New research used genetically engineered plants to produce the sex pheromones of insect pests which can then be used to misdirect and frustrate the pests' mating activities. For the time being, sex pheromones are expensive to synthesise and only cheap enough to be used to protect high-value crops. Hopefully, cheaper processes are on the way.

Field border with wildflowers for pollinators, Essex [Steven Falk]

Making farmland more bio-diverse and more flower-friendly for pollinating insects would be an important additional action. Modern intensive farming commonly involves vast monocultures that demand frequent applications of herbicides and pesticides as well as the eradication of flower-rich hedges, field borders and any other natural or semi-natural areas. In such unfavourable conditions, there are so few flowers that it is a challenge for pollinators to survive at any time of year. Such intensive farming doesn't come close to providing the pollen and nectar resources that bees need. Recent research, led by Dr Lynn Dicks of the University of East Anglia, has shown that 100 hectares of farmland need a bare minimum of two hectares of flower-rich habitat, and a kilometre of flowering hedges, to supply enough pollen to feed the larvae of common bee pollinators. Clearly, the rural environment would benefit from a return to more traditional ecological farming methods, the way land was farmed throughout all but recent farming history. Ecological husbandry improves the environment for pollinators by avoiding extensive monocultures; by mixing in flowering crops other than staples, such as clover, field beans, Borage *Borago officinalis*, sunflowers and Oil-seed Rape, that provide a plentiful supply of nectar and pollen; by keeping patches of natural or semi-natural habitat in the vicinity of crops; and by retaining hedgerows and field borders with abundant flowers. Ecological farming also takes advantage of the free ecosystem services that are present in a healthy environment and provided by wild organisms—free services that include maintaining soil fertility, prevention of soil erosion, control of water flow, natural regulation of insect pests, and pollination.

Recent research, led by Dr Alexa Varah at the University of Reading, studied the effectiveness of pollination in agroforestry systems compared with monocultures. Agroforestry is agriculture with trees—a management system in which trees and other woody plants are grown around and among crops or pasture. The research found that agroforestry sites had double the number of solitary bees and hoverflies, and that arable agroforestry sites had 2.4 times more bumblebees than in monocultures. Solitary bee species diversity also increased tenfold at some sites. The increase in pollinators resulted in more effective pollination and the production of up to 4.5 times as many seeds. As pointed out by the lead investigator *"It is ironic that agriculture, which relies so heavily on pollinators, is actually one of the biggest contributors to their decline. Our*

Wildflower strips enhance agriculture for pollinators [rothamsted.ac.uk]

study finally provides some proof that agroforestry is win-win for wild pollinators and for farmers growing crops that need pollinating."

Rotating crops and leaving fields fallow are two other features of ecological farming. It has been recognised for centuries that crop rotation and fallow years help maintain productive soils. Middle Eastern farmers were already practising crop rotation in 6,000 BC. They had no idea why alternating legumes with other crops helped productivity but experience confirmed the value of the practice. In fact, a seven-year agricultural cycle, culminating in a fallow year, is mandated in the Torah for the Land of Israel. All farming activity is forbidden during the seventh year, a practice still observed by some contemporary farmers. There are also numerous references in the Bible to a seven-year agricultural cycle—a *"Sabbath of the Land"*.

Making farming more ecologically relevant, as it was in the past, would undoubtedly help the survival of wild bees and most other pollinators, and in doing so would make pollination more effective and increase crop yields. Ecological farming does not mean that farming has to be less efficient, less productive or less profitable. There is abundant evidence to the contrary, drawn from research carried out in many parts of the world, including the USA, Europe, India, Bangladesh, Indonesia, Kenya and Malawi. A Canadian study on a very large farm concluded that farmers could maximise profits by leaving up to 30% of their land to grow flowers for pollinators, a strategy that saves on the costs of cultivating land, while increasing yields on the remaining 70%. As emphasised by Greenpeace International—*"Ecological Farming ensures healthy farming and healthy food for today and tomorrow, by protecting soil, water and climate, promotes biodiversity, and does not contaminate the environment with chemical inputs or genetic engineering".*

CHAPTER 9

Protected areas and gardens can help

Help for pollinators does not depend just on using better farming methods. The present UK Government has introduced a National Pollinators Strategy, involving cooperation between Government and many other organisations, including large landowners such as the National Trust, the Forestry Commission and the Ministry of Defence. The Strategy aims to support pollinators on farmland, in the countryside, and also in urban and suburban areas, by ensuring that more flowers are available everywhere throughout the year, and by avoiding unnecessary or excessive use of agricultural chemicals. It is particularly important to manage and restore any areas of natural habitat that still exist. Such areas are often small and isolated and need to be linked by flower-rich corridors, natural or otherwise, such as river valleys, canals, motorway and roadside verges, railway lines, hedgerows and field margins.

Stimulated by the publicity that the pollinator crisis has provoked, other initiatives are at work, many of them involving academics from several universities. The AgriLand project, funded by the UK Insect Pollinators Initiative, surveyed wildflowers across the country to determine which ones are best for pollen and nectar production both as individual plants and on a habitat scale. The top 10 flowers for pollinators were Marsh Thistle *Cirsium palustre*, Grey Willow *Salix cinerea*, Common Knapweed *Centaurea nigra*, Bell Heather *Erica cinerea*, Common Comfrey *Symphytum officinale*, Spear Thistle *Cirsium vulgare*, Common Ragwort *Jacobaea vulgaris*, Hogweed *Heracleum sphondylium*, Bugloss *Lycopsis arvensis* and Chives *Allium schoenoprasum*. Several of these plants are usually regarded as invasive weeds but are also among those recommended by us. The study also found that the most valuable habitats for pollinators are calcareous grassland followed by broadleaf woodland, both habitats that are now very rare in the UK. Arable habitats are the worst for pollinators (out of 11 categories), behind even conifers and bog.

The top 10 flowers for pollinators in the UK [all PC]

Flower meadow, King's College, Cambridge [alumni.cam.ac.uk/image courtesy of the Ely Standard]

In another initiative the Urban Pollinators Project looked for ways to improve pollinator diversity and abundance in cities. Numerous city councils are now creating pollinator-friendly gardens and wildflower meadows in parks, on roundabouts and on central reservations along major roads. Gloucester City Council, for example, (and others) in partnership with the Bee Guardian Foundation, has replaced winter bulbs and bedding plants with wildflowers and saved money in the process; wildflower meadows are being created in several of London's Royal Parks and managed with the help of grazing sheep; and flowers have been planted and are flourishing on at least a few motorway and roadside verges around the country. Recently, in 2020, King's College, Cambridge has joined in this movement. The College's much-photographed, immaculate lawn, sloping down from its famous Chapel to the River Cam, has been transformed into a wildflower meadow, ablaze with poppies and other wildflowers—a 'biodiversity-rich ecosystem', humming with bees, instead of the monoculture of tidy but boring grass that it has been in the past.

Insects and other invertebrates are undoubtedly important. Sir David Attenborough once said—"*If we and the rest of the back-boned animals were to disappear overnight, the rest of the world would get on pretty well. But if the invertebrates were to disappear, the world's ecosystems would collapse.*" The Invertebrate Conservation Trust, known as Buglife, would certainly agree. Buglife is a charity dedicated to protect insects, bugs and invertebrates. To this end, Buglife is promoting the creation of B-lines—flower-rich 'insect pathways' that will run through the countryside, providing good habitat for wildlife, while at the same time linking isolated fragments of existing habitat.

In yet another initiative, the British conservation charity Plantlife has become very active in a campaign to encourage local councils and highways authorities to let neatly mown grass verges become mini meadows where wildflowers and wildlife can flourish—a move that has the potential to save money as well as help pollinators. Plantlife's management guidelines are summed up by the slogan "*cut less, cut later*" so that plants can flourish and have enough time to produce seeds. Plantlife points out that there are over 300,000 miles of rural road in the UK with borders that could be converted into flower-rich habitat for pollinators and other wildlife. Progress is being made. For example, an eight-mile 'river of flowers' along a major route into Rotherham has given much pleasure to road users, as have other roadside meadows in many other cities. According to Plantlife, some drivers still prefer tidy verges but most find that

Roadside planted with wildflowers [Simon Colmer/naturepl.com]

roadside meadows are pleasing to the eye and reduce the stress of driving. One unexpected consequence of the 'lockdowns' caused by the coronavirus pandemic is that local councils have been unable to cut lawns and roadside verges. Wildflowers and wildlife have thrived.

Rooftop gardens are proliferating in towns and cities. The rooftop of Sharrow Primary School in Sheffield has been declared a Local Nature Reserve. And the green roof of the Rolls Royce factory in Sussex, which covers 3.24 hectares (8 acres), has even attracted breeding skylarks. And lapwings and little ringed plovers have attempted to breed on green roofs in Switzerland. Incidentally, green roofs are good for more than just pollinators and birds. They act as insulating layers that significantly cut the cost of air conditioning and heating; they reduce rain run-off; they deliver oxygen and reduce levels of dangerous pollutants; and they give people a sense of wellbeing.

Green roof office, CABI, Wallingford, Oxfordshire [scottbrownrigg.com]

Pollinator-friendly flowers in a Worcestershire garden [ID]

The British are often said to be a nation of gardeners and there is no doubt that urban and suburban gardens can help the plight of pollinators, provided more care is taken to give priority to pollinator-friendly flowers. There are 15 million gardens, amounting to over 400,000 hectares, scattered across the UK—an area equivalent to the combined area of the Norfolk Broads with Dartmoor, Exmoor and Lake District National Parks. It is encouraging that more and more urban and suburban gardens are indeed becoming specialised pollinator gardens, planted with a mix of insect-friendly flowers. Such gardens are flower-rich oases, providing nectar and pollen, breeding sites (sometimes including 'bee hotels') and green stepping-stones that help link scattered fragments of other useful habitat. In some areas the changes are so impressive that cities, suburbia and villages are now often better for pollinators than much of the surrounding, intensively farmed countryside. For example, research by biologists at Royal Holloway College (published in *Proceedings of the Royal Society B*) has shown that bumblebee colonies in urban areas were more successful than colonies in agricultural areas. Significantly, city colonies were more likely to rear future queens and males and had many more workers. Several reasons are probably linked to the greater success of the urban bumblebee colonies—access to plentiful flowers in parks and gardens; less exposure to pesticides than the colonies in farmland; and perhaps less exposure to kleptoparasitic cuckoo bees.

On the whole, native wildflowers are better adapted to the British climate than exotics. In our own garden we prefer to give priority to native plants but we also include some non-natives from other temperate regions of the world, especially a few exceptional species, some now naturalised, such as buddleia and lavender. What matters to pollinators is the accessibility and abundance of nectar and pollen. If an introduced plant offers more easily collected nectar or pollen than the natives around it, pollinators will love it. On the other hand, we avoid flowers, such as red-hot pokers, cannas, and fuchsias, that have evolved to attract hummingbirds, sunbirds or other pollinators that do not exist in the UK.

It is helpful to choose flowers in a variety of shapes and colours to encourage a diversity of pollinators. One size does not fit all. British wild bees—the most important pollinators—are very variable in size, so blossoms should also vary. And it is important to ensure that flowers are blooming through much of the year, especially in early spring. Newly emerged bumblebee queens need early flowers to provide the nectar and pollen resources that they need to start their colonies and feed their larvae. Choices worth considering include Grey Willow and Goat Willow *Salix caprea*, Blackthorn, apples, Winter Aconite *Eranthis hyemalis*, Primrose *Primula vulgaris*, hellebores *Helleborus* spp., Spring Crocus *Crocus vernus* and Rosemary *Salvia rosmarinus*.

A reliable, long-term supply of flowers is also gratifying for gardeners because it will attract and support many predatory and parasitic insects (many of them also pollinators), such as wasps, lacewings, tachinid flies, hoverflies and ladybird beetles, that help to control insect pests. For example, it has been estimated that British social wasps, including the much maligned Common Wasp *Vespula vulgaris* and closely related species, predate 14 million kilograms of caterpillars and other insects in the course of a single summer. And the larvae of about 40% of British hoverflies (including the common Marmalade Hoverfly *Episyrphus balteatus* and many of the other common, black-and yellow, 'typical' hoverflies) are doubly useful—as well as being good pollinators, they have larvae that prey on sap-sucking insect pests, including aphids and thrips. Minute chalcid and trichogrammatid wasps, that lay their eggs in the eggs or larvae of other insects, are also useful. Many have been used for biological control of agricultural pests.

It is very important to avoid the wrong plants. Far too many gardens are stocked with bedding plants (such as pelargoniums, begonias, petunias, pansies and busy-lizzies) and varieties that are useless or near useless to pollinators. Pelargoniums (often incorrectly called geraniums), for example, attract virtually no pollinators. It is also advisable to avoid cultivars with pom-pom or double-flowered varieties that have been so drastically modified by breeding that they lack rewards or are so deformed that it is difficult for pollinators to enter the flower and reach its pollen and nectar. Popular flowers, such as Buddleia *Buddleja davidii*, lavenders *Lavandula* spp. and Wild Marjoram *Origanum vulgare*, attract 100 times as many bees and other pollinators as pelargoniums and other varieties that are unattractive to pollinators.

Plant breeders are mostly concerned with appearance—the size, shape, and colour of the varieties they breed—and care little for the quantity and quality of their pollen and nectar. And many varieties lack the scent that attracts appropriate pollinators. Anyone determined to grow cultivars should try to obtain varieties that are good for pollinators. They do exist. We mention a few in Part 5 and many others are mentioned online. However, the natural versions of flowers are often just as attractive as over-bred varieties and usually more attractive to insects.

Another problem—too many gardens include large areas of unproductive lawn. Even worse are the low-maintenance gardens that have dreary areas of gravel, paving, decking, or even fake turf. Many gardens are overly tidy. Why not allow a little disorder or, better still, a lot? Go for a natural look. And why not convert areas of boring or sterile lawn into a flower-rich meadow that doesn't have to be mowed every week? You don't have to do much—just stop mowing and you will gradually accumulate plenty of pollinator-friendly flowers. And don't bother to clear away dead leaves—leave them for the worms to drag underground and so enrich your soil.

Gardeners genuinely interested in attracting pollinators should consider changing their attitude to so-called weeds. Even the Royal Horticultural Society has lists of weeds that include such attractive species as Creeping Buttercup *Ranunculus repens*, Lesser Celandine *Ficaria verna*, Creeping Cinquefoil *Potentilla reptans*, speedwells *Veronica* spp., Green Alkanet *Pentaglottis sempervirens*, Rosebay Willowherb *Chamaenerion angustifolium* and Dandelion *Taraxacum officinale* agg., most of which are good for pollinators. And other notorious weeds, such as thistles *Carduus* and *Cirsium* spp., sowthistles *Sonchus* spp., knapweeds *Centaurea* spp. and ragworts *Jacobaea* spp., are even better. Remember, too, that many weeds are essential food plants for the larvae of pollinators. A few Stinging Nettles *Urtica dioica*, for example, should be encouraged (or at least tolerated) for they are the most important food plant of the caterpillars of several of our most gorgeous butterflies, notably Peacock *Aglais io*, Small Tortoiseshell *Aglais urticae*, Red Admiral *Vanessa atalanta* and Comma *Polygonia c-album*.

Personally, we like weeds. For years our lawn was pleasingly colourful—full of Daisies *Bellis perennis*, Dandelions, Creeping Cinquefoil, Germander Speedwell *Veronica chamaedrys*, dog-violets *Viola* spp., the occasional thistle and over 150 Bee Orchids *Ophrys apifera*. We were

Corncockle *Agrostemma githago*, Cornflower *Centaurea cyanus* and Pheasant's-eye *Adonis annua* – almost extinct in the wild due to intensive agriculture

amused to read that our much-enjoyed display of wildflowers, including the Bee Orchids, is symptomatic of a "neglected" or "unkempt" lawn. Recently, we further "downgraded" our lawn by converting it into an unkempt perennial meadow. It is even more rewarding and immediately attracted more breeding butterflies, including Meadow Browns *Maniola jurtina*, Gatekeepers *Pyronia tithonus*, Common Blues *Polyommatus icarus* and skippers (Hesperiidae). Reluctantly, we acknowledge the need for unblemished, mowed, green grass at Wimbledon or Lords but fail to see its value in a garden. How can anyone not welcome a colourful sprinkling of violets, speedwells, cinquefoils, Daisies and, yes, even Dandelions? As aptly stated by author Andrew Mason *"If dandelions were hard to grow, they would be most welcome on any lawn"* and William Henry Hudson wrote *"I am not a lover of lawns; ..... the least interesting adjuncts of the country-house. ..... Rather I would see daisies in their thousands, ground ivy, hawkweed, and even the hated plantain with tall stems, and dandelions with splendid flowers and fairy down, than the too-well-tended lawn."*

It is also good to remember that native weed species contribute hugely to the overall diversity of the British flora and fauna. They provide nectar and pollen for pollinators and other insects; many are food plants for the larvae of butterflies, moths and other insects; many more provide fruits and seeds for birds and other animals; and some contribute less obviously by fixing nitrogen or stabilising soils. Eliminating weeds severely reduces overall diversity and has adverse knock-on effects for wildlife. It is a great shame that intensive agriculture in the UK has resulted in the near-extinction or severe decline in the wild of such attractive flowers as Corncockle *Agrostemma githago*, Cornflower *Centaurea cyanus*, Corn Buttercup *Ranunculus arvensis*, Pheasant's-eye *Adonis annua* and Corn Marigold *Glebionis segetum*, most of them once considered as pestilential weeds.

Although many weeds are excellent for pollinators, there is no doubt that a few of the worst introduced, alien plants are an undeniable scourge in the countryside and gardens. The worst are Japanese Knotweed *Reynoutria japonica*, Rhododendron *Rhododendron ponticum* and Himalayan Balsam *Impatiens glandulifera*. All deserve to be eradicated. All were introduced to the UK by gardeners.

Which brings us to garden services—companies that offer to create and maintain weed-free, perfect lawns, usually with the liberal use of weed killers and other chemicals. Their lists of common pernicious weeds found in the UK include many of the same pollinator-friendly species listed as weeds by the Royal Horticultural Society. They even include White Clover *Trifolium repens*, Red Clover *Trifolium pratense*, Common Bird's-foot-trefoil *Lotus corniculatus* and other legumes—all good for pollinators as well as being nitrogen fixers that improve the health and fertility of soil. Luckily, there are many wildlife gardeners who like so-called weeds as much as we do. We often exchange weeds with our friends.

PART 4
The natural history of pollination

Small Sallow Mining Bee *Andrena praecox* on Grey Willow *Salix cinerea* [PC]

CHAPTER 10
Flower structure, rewards and advertisement

If the ongoing pollinator crisis in the UK is to be reversed, it would be useful for all those interested in helping, particularly gardeners, to be familiar with the natural history of flowers and their pollinators. In practice, people sometimes have surprisingly little appreciation of the basic purpose of flowers. Flowers have a clear-cut function—namely reproduction. This begins with sex (pollination)—the transfer of the male genetic material, that is pollen, from the anthers of male flowers to receptive female stigmas. Some flowers are pollinated by wind or water but most are designed to broadcast signals that attract pollinators—insects, birds or mammals— and bribe them (or trick them) into carrying their pollen from flower to flower. In practice, pollinators function as sexual go-betweens for plants and their flowers. With attractive colours, alluring scents and rewards of pollen and nectar, flowers attract pollinators and induce them to disperse their pollen from one to another. The structure of flowers makes sure that visiting insects or other animals—potential pollinators—brush against their anthers and stigma, making the successful movement of pollen from one to the other more likely. However, it should not be thought that pollination is the final objective. It is just the first step in a process that leads on to fertilisation and ends with the production and dispersal of seeds. Once deposited on the stigma, grains of pollen absorb moisture, usually in the form of stigmatic secretions, and then germinate. A pollen-tube results and grows down the style and fertilises an ovule. Once fertilised, ovules develop into seeds that, if successfully dispersed, bring into being a new generation of plants.

The structure of flowers

Flowers are dedicated reproductive structures whose ultimate function is to capture wind-blown pollen or attract pollinators, then produce seeds and finally facilitate seed dispersal. Typically, the basic structure of a flower has four sorts of flower parts—sepals, petals, stamens and carpels—arranged in separate, concentric whorls. The outermost whorl consists of sepals that are often green and leaf-like. Together they form the calyx which encloses and protects the developing flower bud (though sometimes the sepals resemble colourful petals). Next comes a whorl of petals, which together form the corolla. The petals are generally colourful, providing a visual signal that advertises the flower to pollinators. Both the sepals and petals are sterile and, once they have accomplished the job of attracting pollinators, are not involved in the actual process of making seeds. It is only the stamens and carpels that are involved in the production of seeds. First, there are usually a large number of stamens, arranged in one or more whorls, each consisting of a filament topped by an anther that opens to release pollen. Finally, there is central group of carpels, each with a receptive stigma at its tip and each containing an ovule or ovules. After being fertilised, the ovules develop into seeds.

The structure of flowers is very variable. A few examples are mentioned below and others are described in the species accounts in chapter 14 and in the captions of photographs. The sepals, petals, stamens and carpels vary in number. Many dicot flowers, such as buttercups, crane's-bills and primroses, have five sepals and five petals, whereas most monocots, such as lilies and irises, have only three of each, virtually the same in shape and colour and both contributing equally to the visual display. Sepals and petals are then called tepals. There are many other variations among the dicots. Crucifers have four sepals and petals with the latter arranged in a cross—hence the name crucifers. Most water-lilies have more than 20 petals. In some members of the buttercup family, including Winter Aconite *Eranthis hyemalis*, hellebores,

Borage *Borago officinalis* showing petals and sepals

Spring Crocus *Crocus vernus* with three petals and three near identical sepals

Bumblebee Hoverfly *Volucella bombylans* at Gum Rock-rose *Cistus ladanifer*

Marsh-marigold *Caltha palustris* with colourful petal-like sepals

Marsh-marigold *Caltha palustris*, Globeflower *Trollius europaeus*, and many anemones and pasqueflowers, the petals are much reduced or missing altogether. It is their sepals that are brightly coloured and petal-like, providing the visual signal that attracts pollinators.

Flower structure also differs in its symmetry. Many familiar flowers, such as buttercups, poppies, crane's-bills, mallows and rock-roses, are bowl or saucer-shaped, with radial symmetry

Common Carder Bumblebee *Bombus pascuorum* at Viper's-bugloss *Echium vulgare*

Hummingbird Hawk-moth *Macroglossum stellatarum* at Buddleia *Buddleja davidii* (PC)

Perennial Sunflower *Helianthus × laetiflorus*: each petal is an individual flower

that allows pollinators to land on them facing any direction. Many are 'open access' flowers that are pollinated by diverse, small generalist pollinators with a short tongue. Other flowers with radial symmetry, such as bellflowers, bindweeds and heaths, have sepals and petals that are fused together to varying degrees to form bell-shaped or tubular flowers. Most of these are pollinated by insects with a longish tongue.

There are also many flowers with bilateral symmetry, including legumes, dead-nettles, snapdragons, toadflaxes and orchids. Such flowers can be divided down the middle into two mirror-image halves. Their more complicated structure is a barrier to many insects, so most tend to be visited and pollinated by more specialised insects. Their structure, especially the shape and location of a place to land, obliges insects to orientate and adopt postures that increase the chances of successful pollination taking place. Flowers with both sorts of symmetry are sometimes pendant, a feature that deters or blocks many unwanted insect arrivals by demanding more expertise in their flying skills.

Flowers also differ by occurring singly in some species or by being clustered together in inflorescences that make them more conspicuous and attractive to pollinators. Many familiar flowers are clustered together in umbels or spikes in which the flowers are sometimes very close together, as they are in umbellifers, Red Valerian *Centranthus ruber* and Buddleia *Buddleja davidii*, but often further apart, as in Viper's-bugloss *Echium vulgare*, foxgloves and hollyhocks. There are other types of inflorescence in which the individual flowers are so small and packed so tightly together that they look like a single flower. Common, familiar examples include clovers and daisies. In reality, a clover is a collection of many flowers clustered together at the end of the stalk. And a daisy is a collection of tiny flowers known as florets. The 'petals' of the common Daisy *Bellis perennis* and relatives are ray florets that look like regular petals. And the central disc of a Daisy is a dense group of tube florets that produce nectar. All members of the daisy family have compact inflorescences but not all have the same structure. Species that resemble dandelions differ in having only petal-like, ray florets, while thistles and their relatives are clusters of tube florets. Although they are inflorescences, both daisies and thistles look like single flowers and that is how they are treated by both pollinators and most people.

There are also numerous flowers, particularly in temperate areas, that are pollinated by wind or water rather than insects or other animals. They are not trying to attract insects and are very different from insect pollinated flowers in their structure and appearance. They lack colourful petals, scents and rewards and have structural modifications that aid the dispersal and capture of wind-blown or water-borne pollen. Grasses and many common trees—e.g. oaks, elms, alders, and hazels—have wind-pollinated flowers. At first sight, most do not resemble insect-pollinated flowers in any obvious way.

Rewards

Flowers attract pollinators by offering a reward, usually food in the form of pollen and/or nectar. Pollen and nectar are by far the commonest rewards but there are others that are more esoteric, most of them more common in tropical habitats but relatively rare in the UK and other temperate regions.

Pollen

Pollen was probably the main reward available to the first pollinators, thought to be beetles, flies, moths, and thrips. The fossil record (based largely on insects fossilised in amber) shows that the first pollinating insects were active as long as 200 million years ago, in other words about 70 million years before flowering plants first evolved. At that time, they were attracted to trees and other plants by stigmatic secretions and/or the pollen of conifers and relatives that are now extinct. Nowadays, in the age of flowering plants, pollen is still important, serving as an essential reward for bees and a few other insects (as well as numerous tropical bats). Pollen is very nutritious, containing protein, carbohydrates, amino acids, minerals and vitamins.

A male Red-tailed Bumblebee *Bombus lapidarius* with pollen grains

Because of its high nutrient content, pollen is expensive for plants to make. Nevertheless, pollen is the only reward provided by many flowers that lack nectar. Worldwide, there are at least 20,000 species of flowers in which pollen is the only reward. They have pollen that is dry and powdery, not as sticky as typical pollen, and available in greater amounts than is usual in other flowers. The vast majority of pollen-only flowers are bee-flowers. The bees collect the pollen methodically and pack it into pollen baskets on their legs, or other storage organs, before transporting it to their nests. Pollen grains are also trapped passively in the coats of hairy bees and are available to pollinate other flowers. Later, bees gather any remaining grains by grooming. In the UK poppies and rock-roses are common pollen-only flowers in the countryside or gardens and attract many visits by bees.

Most other pollen-only flowers, particularly in the tropics, have to be buzz-pollinated (more accurately called vibratory pollen collection). British buzz-pollinated flowers include Borage *Borago officinalis* and species of *Solanum*, including Woody Nightshade and varieties of tomatoes. Buzz-pollination is used by bumblebees and numerous solitary bees in several families, but not by Honey Bees. Flowers that are buzz-pollinated have tube-like anthers that expel pollen that is dry and dust-like, fine enough to escape from a tiny pore at the tip of the anther. Bees grasp the flowers with their legs or mandibles and vibrate their thoracic wing muscles rapidly at frequencies of around 320–410 cycles per second. In doing so, they transmit vibrations through their body to the flower's anthers, causing pollen to spurt from the pores and dust the visiting bee. Buzz pollination can be simulated using a tuning fork that mimics the vibrations of a bee. A 512 Hz tuning fork works well. The process of buzz pollination is probably helped by an electrostatic charge carried by the bees on their hairy bodies. Pollen literally jumps a few millimetres onto a bee and gets caught amongst its hairs. Bees then groom themselves, collect the pollen, and pack it into a storage organ. Without the electrostatic charge, much pollen would be lost to gravity. Buzz pollination ensures that pollen is collected only by insects that vibrate at appropriate frequencies.

Common Carder Bumblebee *Bombus pascuorum* at Borage *Borago officinalis,* a buzz-pollinated flower

Tree Bumblebee *Bombus hypnorum* collecting nectar at Hollyhock *Alcea rosea* flower

Nectar

Nectar is the reward used by most British flowers to attract insect pollinators. It is also used by numerous tropical flowers to attract nectarivorous bats, rodents and birds. Nectar is a dilute solution of sugars, including sucrose, fructose and glucose. It is secreted by nectaries—specialised flower tissues that sometimes reveal themselves by tiny drops of exuded nectar. Disc nectaries form an obvious ring surrounding the base of the ovary and stamens. An interesting variation is found in so-called 'revolver flowers' in which a visiting pollinator is confronted by a ring of narrow tubes (resembling the ammunition chambers of a revolver) that give access to nectar. An insect must circle and probe all the access tubes to obtain all the available reward. Classic British examples include bindweeds and hollyhocks. Nectar can be secreted in many other places, including sepals, petals, carpels, and stamens. In some flowers, nectar is stored in spurs. British native or garden flowers with elongated spurs include toadflaxes, nasturtiums, columbines, violets and some orchids. Such flowers can only be pollinated legitimately by long-tongued bees, butterflies and moths but are often robbed by short-tongued bumblebees that reach the nectar by biting holes in the spurs.

Although nectar is a solution of sugars—an excellent source of energy—it is less good than pollen as a source of other important nutrients, such as amino-acids, proteins and minerals. Nectar is also important because it provides a crucial supply of water for insects during dry weather in arid environments. The amount and concentration of the nectar provided by different flowers is variable and intended to encourage frequent visits by the most suitable pollinators. Flowers pollinated by bees and other insects secrete relatively small quantities of concentrated nectar while flowers pollinated by tropical bats and hummingbirds produce copious, rather dilute nectar.

Miscellaneous rewards

Other less common rewards offered to pollinators include stigmatic secretions, floral oils and resins, sterile food stamens, perfumes and breeding sites.

The secretions exuded by stigmas are primarily involved in keeping the stigma moist, but also aid the capture of pollen grains, and their subsequent germination. In addition, stigmatic secretions sometimes serve as a sugary reward that is attractive to tiny visiting pollinators, mainly flies and thrips. It is thought likely that stigmatic secretions were an important reward for some of the insects that were pollinators of the very first flowers.

Flowers that use oils as a reward occur in several plant families. The oils are a rich source of energy and are secreted by specialised glandular areas on the petals. There are many oil-flowers in the tropics but only one in the UK—Yellow Loosestrife *Lysimachia vulgaris* which produces oil that is collected, together with pollen, by female Yellow Loosestrife Bees *Macropis europaea*. The bees use the oil to both provision and waterproof their nests which are often constructed in damp ground subject to flooding.

In the tropics, though not in the UK, there are also plants that secrete resins. These resins are collected by stingless bees and some solitary bees for use in constructing their nests. The resins are often collected as they seep from wounds in tree trunks and branches. However, there are also flowers in two genera of plants—*Dalechampia* (Euphorbiaceae) and *Clusia* (Clusiaceae)—that produce floral resins that are collected by stingless bees. As well as being a first class resource for building nests, the floral resins are said to have bactericidal qualities that protect the eggs and larvae being reared in the bees' nests.

In the tropics, rewards are much more diverse than in the UK. There are flowers that offer sterile stamens as edible rewards. There are orchids that attract their bee pollinators with perfumes that male bees collect and use to attract females. And in the Neotropics, there are flowers in the daisy and borage families that attract male glasswing butterflies (Ithomiinae) with alkaloids that the male butterflies use to make pheromones to attract females and for chemical defence. There are also tropical flowers in several families that attract beetles and flies by providing places that appear to be suitable to lay their eggs. Sometimes the places really are suitable but others are often fake. These rare and unusual enticements cater for only a small minority of pollinators with special needs. Even so, they are vital for the species involved, all of them participants in extraordinary pollination partnerships, mostly unmatched in temperate areas.

Yellow Loosestrife Bee *Macropis europaea* collecting oil from Yellow Loosestife *Lysimachia vulgaris* [NO]

A stingless bee in the tropics collecting resin from a *Clusia* flower

Warmth

Warmth is a less obvious reward but one that many flowers offer in areas with a cool climate. In the northern temperate zone there are many bowl-shaped flowers that function like parabolic reflectors, achieving higher than ambient air temperatures by reflecting heat to their centre. Other flowers achieve a similar result with blackish blotches that function much like solar panels. Some flowers absorb even more heat by tracking the sun's movement so that they face towards it at all times. This extra warmth is likely to be advantageous to both the flowers and their pollinators. The plant benefits because extra warmth helps seed development in any plants that flower in

A hoverfly *Anasimyia lineata* and micro-moths attracted to a buttercup's bowl-shaped flower [PC]

early spring. And the pollinators benefit from the additional heat that enables them to keep warm and remain active even in early spring in northern latitudes or on high mountains. Examples of bowl-shaped flowers that provide warmth as a reward include poppies, anemones, pasqueflowers and buttercups.

Advertisement

Pollinators recognise preferred flowers by a combination of colour, shape, and scent. In fact, flowers have been described as 'sensory billboards' that send 'come hither' signals to pollinators. Most are visual or olfactory signals , though there are plants in Central and South America that attract nectar-eating bats using 'sonar guides'. These flowers have modified petals that work like a parabolic reflector and send the bats' echo-locating calls back towards them.

Colour

Most pollinators have good colour vision so flowers use colour to advertise the presence of rewards. The colours used are determined by the visual abilities of the animals they are trying to attract, which are usually very different in mammals, birds and insects. In temperate Britain there are none of the specialised mammal or bird pollinators—bats, hummingbirds, sunbirds, etc.—that are common in tropical regions. British pollinators, all insects, show some colour preferences but there are also considerable overlaps. Bees distinguish colours towards the blue end of the spectrum, including ultra-violet (which we and most other mammals cannot see), but not red, so it is no surprise that bee-pollinated flowers are usually some shade of ultra-violet, blue, yellow or white. Most flies prefer blue, cream, white or yellow-green; butterflies are attracted by red, pink, purple, yellow or white; and nocturnal moths by cream or white. However, even though most pollinators have innate colour preferences, they are also capable of learning quickly that other colours, or even flower shapes, are associated with rewards. Many pollinators sometimes visit almost any flower.

Nectar guides

Nectar guides are a conspicuous feature of many flowers. The German theologian and teacher, Christian Sprengel, was the first naturalist to understand and describe the significance of nectar guides. Sprengel realised that the guides direct flower visitors towards nectar or other rewards and it was the yellow ring around the centre of forget-me-not flowers that originally attracted his curiosity. Lots of flowers are now known to display nectar guides and the guide patterns come in many forms—bull's-eye patterns in contrasting colours, converging lines,

White-tailed Bumblebee *Bombus lucorum* at a globe-thistle *Echinops* sp.

Common Mallow *Malva sylvestris* flowers with nectar guides [PC] arrow-like pointers and blotches of contrasting colour. Some nectar guides, including guides found in sunflowers, dandelions, borage and evening-primroses, are revealed only in ultraviolet light. Most insect pollinators can see ultra-violet guide patterns but they are invisible to humans and other vertebrates. There are also some nectar guides that change colour soon after a flower has been pollinated and is no longer the source of a reward. For example, the attractive, yellow ring surrounding the centre of forget-me-nots changes to an unattractive dingy white.

Scent

The majority of pollinators, with the notable exception of birds, have a highly sensitive sense of smell and live in environments in which they are surrounded by odours—pleasant or foul. Flowers take advantage of this aptitude by using smells, in addition to colour and directional patterns, to advertise to pollinators. To our unrefined human noses, the scents of flowers belong to one or other of just four major groups. Those in the first group have the enchanting fragrances characteristic of roses, freesias, gardenias, frangipani, tobacco flowers, hyacinths and many other flowers. These scents are typical of flowers that attract bees, butterflies and moths. Another group includes the often mildly unpleasant yeasty or fruity odours characteristic of fermenting fruit. They are usually found in tropical flowers that are visited and pollinated by bats or beetles, or sometimes by flies. The third group caters for the putrid stenches that appeal to blowflies, flesh flies and dung beetles—insects that lay their eggs in rotting carcasses or excrement. And the final group is restricted to the odours found in the orchids that mimic the sexual pheromones used by some male bees and wasps to attract mates.

The scent of some flowers 'switches off' soon after they have been pollinated, letting pollinators know that rewards have been used up and are no longer available. Flower scents also change if and when visiting pollinators depart, leaving a short-lived odour 'footprint' or message indicating that nectar or other rewards are no longer available. This sort of scent-marking is very common in bees and the 'footprint' of some bees is often recognised by other bee species.

The foul smell of the flowers of Carrion Plant *Hoodia* sp. attracts blowflies, etc.

CHAPTER 11
Self-pollination and cross-pollination

Flowers are self-pollinated whenever a flower pollinates itself with its own pollen or pollinates other flowers on the same plant. Flowers are cross-pollinated whenever pollen is carried between flowers on separate plants.

Bisexual flowers and self-pollination

The flowers of most plants are both male and female (called bisexual, hermaphrodite or perfect flowers). Bisexual flowers have functional anthers and a functional stigma in such close proximity that self-pollination, followed by seed production, is possible or highly likely. Self-pollination is cheaper than cross-pollination because the plants involved use less resources. Compared with close relatives, they generally have flowers that are both smaller and fewer in number, and each produces less pollen and little or no nectar. On the other hand, self-pollination can be disadvantageous because it results in progeny without the genetic diversity conferred by cross-pollination. This genetic diversity is potentially beneficial, especially if habitats are altering (e.g. because of climate change).

Nevertheless, self-pollination can be useful in areas where pollinators are scarce or absent, as they often are in the case of ephemeral species and in others that flower in regions where pollinators are scarce because the weather is unpredictable and often cold or wet; and also in pioneer plants that are adapted to disperse and colonise new geographical territory. Self-pollination is generally rare in perennial plants that live for a long time.

Self-pollination is also beneficial as a back-up system if cross-pollination fails—seeds resulting from self-pollination are obviously better than none at all. Bee Orchids *Ophrys apifera*

Bee Orchid *Ophrys apifera* – self-pollination

in the UK and elsewhere in northern Europe provide a useful instance of self-pollination used as a back-up. In southern Europe, Bee Orchids are rarely self-pollinated because most manage to attract male long-horned bees (*Eucera* and *Tetralonia*). The orchids are then pollinated by 'pseudocopulation' (i.e. when a male bee is fooled into attempting to copulate with a flower that mimics some features of a female bee well enough to elicit a mating response). However, the relevant species of bees are absent or very scarce in the UK (and elsewhere in northern Europe), areas where self-pollination is the norm. Soon after a British Bee Orchid opens, its pollinia swing away from the anther but remain hanging close to the stigma. Even a gentle breath of wind is usually enough to blow the pollinia onto the stigma, so that self-pollination results.

Bisexual flowers and cross-pollination

It is obviously better for flowers to be self-pollinated, rather than not pollinated at all, but cross-pollination is almost always preferable because it has the potential to spread beneficial genes in novel combinations that might be advantageous to the plant. This is why most flowers have adaptations that encourage cross-pollination and make self-pollination difficult or even impossible. Three sorts of adaptations are found—first, flowers can avoid self-pollination by being self-incompatible; secondly, by having anthers and a stigma that are separated by a short distance or sometimes by being on different plants; or thirdly by having anthers that release their pollen either before or after the stigma is mature and receptive, definitely not at the same time.

Obviously, if plants are self-incompatible, they must be able to distinguish their own pollen from the pollen of other plants. The ability to distinguish 'self' and 'non-self' is quite common in plants and animals, though it is almost always rejection of 'non-self' that is most important (as it is, for example, in the human immune system). However, in self-incompatible plants it is the rejection of 'self' that occurs. It can even involve the rejection of pollen from closely related plants. For example, in the Common Poppy it has been shown that any one plant can interbreed with around 80% of other nearby plants but not with the rest, the majority of which are its closest relatives.

Spatial incompatibility is found in plants with flowers in which the stigmas and anthers are significantly far apart. In many, including poppies, Marsh-marigolds, buttercups and crane's-bills, the stamens are simply splayed away from the stigma. Arriving pollinators usually land in the centre of the flower so, if they are carrying pollen, it is very likely to be deposited on the central stigma. Visitors typically come into contact with the anthers only afterwards, as they crawl among the stamens collecting pollen, after which they depart for another flower, carrying a fresh load of pollen.

A very distinctive version of spatial incompatibility is found in flowers that have two or three different versions with styles of different lengths (known as heterostyly). Familiar British flowers with these adaptations include Primrose *Primula vulgaris* and Cowslip *Primula veris*, Wood-sorrel *Oxalis acetosella* and Purple-loosestrife *Lythrum salicaria*. Primroses and Cowslips each have two forms in which the anthers and stigmas are at alternate positions in flowers in different plants. In the 'pin' form, the style is long enough for the stigma to show at the opening to the corolla, while the anthers are situated only about half-way up the corolla tube. In the 'thrum' form the positions of the stigma and anthers are the other way around. As a result, pollen is deposited on two distinctly separate places on the long tongue of visiting pollinators, such as Dark-edged Bee-flies *Bombylius major* and snout hoverflies *Rhingia* spp. It follows that pollen collected from pin flowers will be accurately transferred to thrum stigmas, and vice versa. Some other plants, such as Wood-sorrel and Purple-loosestrife differ in having flowers with three different style lengths—short, medium and long—coinciding with three different anther positions.

Spatial incompatibility is also found in plants that are dioecious i.e. plants that have unisexual flowers with male and female flowers separated by being on different plants (see p. 56). Because self pollination is clearly not possible in dioecious flowers, they depend on being cross-pollinated by an external agent—i.e. by an animal or wind or water.

Common Poppies *Papaver rhoeas* cannot interbreed with close relatives

Foxglove *Digitalis purpurea* – temporal incompatability

Marsh-marigold *Caltha palustris* – spatial incompatability

Temporal incompatibility occurs when a flower cannot self-pollinate because its anthers and stigma are not mature and functional at the same time. There are two versions of temporal incompatibility—in the first version a flower's anthers are mature and shed pollen before its stigma has become receptive, while in the second version a flower's stigma is receptive before its anthers are ready to shed pollen. 'Anthers before stigma' is the commoner form and is, for example, the norm in the daisy family. It is also characteristic of plants with upright flowering spikes, such as claries *Salvia* spp., mulleins *Verbascum* spp. and Foxgloves *Digitalis purpurea*.

The other form, 'stigma before anthers', is less common than 'anthers before stigma' but occurs in several familiar flowers, including buttercups, crucifers, and in some wind-pollinated flowers, such as plantains. In some of these plants, if they have not already been pollinated, the stigma sometimes remains receptive until the anthers mature and release their pollen. So, self-pollination is sometimes possible but only provided prior cross-pollination has failed.

Red Valerian *Centranthus ruber* is interesting because it illustrates both spatial and temporal incompatibility. Young flowers have a single stamen that sheds pollen while the style remains short and immature. Subsequently, the stamen reflexes and twists behind the flower, well out of the way, while the style elongates and becomes receptive. Of course, neither spatial nor temporal incompatibility prevents self-pollination with pollen from another genetically

Hazel *Corylus avellana* male catkin shedding poillen and female flower showing red strigmas

identical flower on the same plant. How likely this is to happen depends on how many flowers are open on the plant at the same time and on the amount of pollen remaining on a pollinator as it moves from flower to flower.

Unisexual flowers

Worldwide about 85% of angiosperms have bisexual flowers, and the remaining 15% have flowers that are unisexual—they are either male or female. Of the plants with unisexual flowers, about a third (i.e. 5%) are monoecious (a word derived from the Greek for 'one house')—in other words flowers of both sexes occur on the same plant. The other unisexual flowers (10%) are dioecious ('two houses')—male and female flowers are segregated on different plants so that flowers on any one plant are either all males or all females.

Monoecy is strongly associated with wind pollination. In the UK, monoecious plants include numerous common wind-pollinated trees, such as the oaks, beeches, hazels, birches, alders, and hornbeams—the trees that dominate British woodland. Their male and female flowers tend to be quite well spaced, though wind-blown pollen sometimes results in self-pollination. This is less likely to happen in species, such as hazels, alders, and others, in which the female flowers are receptive only before the male flowers are ready to release pollen.

Dioecy occurs in no more than 5% of British plants though some, including holly, willows and stinging nettles, are common and familiar. Self-pollination is obviously not possible in dioecious flowers, because male and female flowers are segregated on different plants. These species depend on being pollinated by an external agent, by an insect or wind (water pollination is more or less confined to seagrasses). In the UK, most dioecious flowers are pollinated by insects, rather than wind. The male flowers tend to be larger and more conspicuous than the female flowers which increases the probability that pollinators will visit them first. Males before females is obviously essential if pollination is to be successful. A few dioecious species are wind-pollinated, including poplars, nettles and mercuries.

Asexual reproduction

There are a significant number of plants that do not always have to be pollinated and instead reproduce asexually, a procedure that results in progeny that are genetically identical to the parent plant. The new plants are formed vegetatively but can be derived from various different parts of a plant, including bulbs, rhizomes, and runners. The process can produce clones that occasionally extend over huge areas. There is a famous cloned grove of male quaking aspens in the USA (with the nickname Pando) which has around 47,000 stems growing in an area of 43 hectares. The stems average only about 130 years old though Pando's network of roots is considered to be at least 80,000 years old and some authorities think its age might be closer to 1 million years. Pando has been shown to be a single living organism that weighs about 6,000 metric tons. Pando is thought to be the world's largest organism.

There are also a few plants with flowers capable of producing seeds from unfertilised ovules, a process known as apomixis. This procedure is not uncommon in the daisy and rose families. It is, for example, the most common form of reproduction in dandelions, hawkweeds, whitebeams and brambles.

CHAPTER 12

Syndromes, flower constancy and nectar thieves

Flowers and pollinators

Worldwide, the majority of flowers rely on insects, or less often on birds or mammals, to move their pollen from flower to flower. The flowers involved have characteristic sets of features—their structure, colour, scent, and rewards—that match the sensory capabilities of their usual pollinators, whether they are mammals, birds, bees, butterflies or other insects. These sets of features, usually called 'pollination syndromes', make it possible to foresee a flower's probable pollinator. In the tropics, for example, it is usually easy to predict whether a flower's intended pollinator is a bat, hummingbird, bee, butterfly or beetle. However, in the UK, where relatively unspecialised insect pollinators predominate (and vertebrate pollinators, such as nectarivorous bats, rodents and hummingbirds, are absent), these syndromes are sometimes helpful but much less distinct than in the tropics.

Collared Inca hummingbird at *Capanea* and a nectar bat visiting *Mucuna*

There are also many flowers that are pollinated abiotically—by wind or water—rather than animals. They too have suites of adaptive characters that promote pollination but they are usually very unlike the characteristics of flowers pollinated by insects, birds or mammals. As already noted, they lack colourful petals, scents, and rewards, and many do not resemble flowers in any obvious way. In the past, it was thought that wind pollination is a primitive condition in angiosperms (as it is in conifers and other gymnosperms) and that pollination by insects and other animals evolved much later. However, opinion has now changed. The fossil record is unambiguous in showing that wind pollination is not primitive in flowering plants. The earliest fossil flowers had characters clearly indicating that they were insect-pollinated. Furthermore, wind pollination is still rare in the primitive flowering plant families that still survive today. Contemporary survivors include numerous species of magnolias and water-lilies, all of which are pollinated by insects.

Needless to say, the frequency of occurrence of different pollination syndromes varies enormously in different climatic zones. Pollination by bats, other mammals and birds (particularly by hummingbirds, honeyeaters and sunbirds) is common in humid tropical regions, where flowers are available throughout the year, but much less common, or absent, in regions with prolonged, cold winters and in deserts with long, harsh dry seasons. Wind pollination is rare in tropical rainforests where up to 98% of flowering plants are pollinated by animals. This is because individuals of most tree species occur at very low densities and are usually too far apart for wind pollination to be effective. On the other hand, wind pollination is very common among trees in temperate areas and anywhere else where individuals of many plants grow in relatively close proximity or crowded together in monocultures. In the UK, pollination by animals is restricted to insects and wind pollination is very common.

Flower constancy

Many pollinators visit more than one species of flower as they forage—behaviour that, from a flower's point of view, risks different types of pollen becoming mixed and carried to the wrong destination. On the other hand, there are some pollinators that show 'flower constancy',

Red-tailed Bumblebee *Bombus lapidarius* worker at Cornflower *Centaurea cyanus*

behaviour that improves the effectiveness of cross-pollination. The pollinators involved, including many bees (especially bumblebees), hoverflies and butterflies, often concentrate for a while on just one especially rewarding flower species and completely ignore others. The length of time a pollinator continues to visit just one species of flower is very unpredictable. Visits may be repeated at regular intervals for hours, or perhaps for days if lots of flowers continue to be available. Visits also depend on how speedily the nectar, pollen and other rewards are used up. Flower constancy is learned behaviour on the part of the pollinators and allows them to visit and exploit flowers that differ from their innate preferences. Such constancy benefits the plants concerned because it greatly increases the likelihood of successful cross-pollination. The benefit to pollinators is not so obvious though concentrating on just one type of flower is probably more efficient than having to constantly adjust foraging behaviour to exploit a variety of flowers that differ structurally.

Flowers and nectar thieves

It is important to remember that many of the animals that visit flowers are not pollinators. The association between plants and animals is entirely self-serving and rewards are often stolen by animals that play no part in pollination. Buds, flowers, nectar and pollen are valuable, sought-after foods and often eaten or damaged by diverse mammals and birds, particularly deer, squirrels, pigeons and finches. The impact is sometimes severe. For example, in the 1950s and 1960s, Bullfinches caused so much damage to the flower buds and blossom of fruit trees in British orchards that they were treated as serious pests. In excess of a thousand bullfinches were legally trapped every year in some commercial orchards. Nectar is also stolen by many insects. In the UK, the most common offenders are short-tongued bumblebees that habitually chew holes into the nectaries of any flowers in which the nectar is inaccessible. Other insects that cause damage include beetles that browse on flower tissues, including stigmas and anthers.

If they are to survive and flourish, plants have to arrive at an advantageous balance between encouraging visits by genuine pollinators while at the same time protecting their flowers from damage inflicted by herbivorous visitors. Because of this, many flower species are protected by toxic or foul-tasting chemicals.

Early Bumblebee *Bombus pratorum* stealing nectar (piercing) from a Columbine *Aquilegia* flower

CHAPTER 13

The pollinators

nsects are the only significant animal pollinators in the UK and it is convenient to discuss them in seven groups, each of which has distinctive adaptations for exploiting and pollinating flowers. The seven groups are: opportunistic unspecialised insects; bees; wasps; flies; beetles; butterflies; and moths. A few other insects (e.g. earwigs, cockroaches, bugs, etc.) are occasional flower visitors, sometimes feeding on nectar or pollen, but none are significant pollinators. Wind and water are the only other agencies that are significant pollinators of British plants. Wind pollination is very important but water pollination is almost confined to a few marine grasses.

Pollination by opportunistic, unspecialised insects

Many flowers in many families are small, relatively unspecialised and pollinated by a varied assortment of small, opportunistic insects which have a short tongue. Insects in this category include lacewings, tachinid flies, flesh flies, sawflies, ichneumon wasps, small beetles and thrips.

Many of the plants visited and pollinated by small, unspecialised insects have small, fragrant, white flowers with both nectar and pollen that is easy to get at. These flowers are often gathered together in compact flat-topped inflorescences that provide pollinators with a good platform on which to land. British flowers of this sort include most of the umbellifer family, including Cow Parsley *Anthriscus sylvestris*, chervils *Chaerophyllum* spp., Wild Carrot *Daucus carota*, Hogweed *Heraclium sphondylium*, Hemlock *Conium maculatum*, Alexanders *Smyrnium olusatrum* and many others. Umbellifers are abundant in the UK and most other temperate habitats in both Eurasia and North America. Other plants preferred by small generalist pollinators, include several with umbellifer-like inflorescences, such as Yarrow *Achillea millefolium*, Elder *Sambucus nigra*, Hawthorn *Crataegus monogyna*, Firethorn *Pyracantha coccinea*, Rowan *Sorbus aucuparia*, Whitebeam *Sorbus aria* and Wayfaring-tree *Vibernum lantana*. Other plants that are popular with generalised, opportunistic pollinators include numerous composites (daisies, dandelions, etc.) and many easily accessible, bowl- or saucer-shaped flowers, such as anemones, crane's-bills, wild roses and brambles.

Most umbellifers are very similar in appearance and often difficult to distinguish. It has been proposed that they are an assemblage of Mullerian mimics and that other plants with similar flowers, such as Elder, Hawthorn and Rowan and Yarrow, may well belong to the same mimicry complex. It is presumed that this convergence of floral characteristics improves the chances of pollination. It is very likely that Mullerian mimicry is also involved in flowers in the daisy family, in which numerous species have yellow or orange flowers that are so similar that they can be difficult to distinguish even by us (e.g. dandelions, hawk's-beards, hawkbits and hawkweeds).

Small, unspecialised pollinators also include thrips (Thysanoptera), commonly known as thunder-flies and notorious for appearing behind the glass in picture frames. Worldwide, about 6,000 species have been described. Thrips are minute insects. Most are no more than 1–2 mm long though some of the largest predatory species reach 14 mm long. Thrips have piercing and sucking mouthparts and most species feed on plant tissues—buds, flowers and tender young leaves—and many are garden pests, causing discoloration to foliage and deformities to developing vegetables and fruit. Some are also known to affect crops by transmitting viruses. There are also many species that live in flowers, where they feed on pollen, and a few that are significant pollinators. In areas in the far north, where conventional pollinators are often rare, thrips are the major pollinators of ling and other types of heather. Elsewhere in the world, thrips are important pollinators of the huge, mass-flowering dipterocarp trees in the rainforests of South East Asia.

Common Red Soldier Beetles *Rhagonycha fulva* and other small pollinators at Wild Carrot *Daucus carota*

A small hoverfly *Syritta pipiens* at Yarrow *Achillea millefolium*

A tachinid fly *Eriothrix rufomaculata* at Common Ragwort *Jacobaea vulgaris*

Pollination by bees

Bees are by far the most important pollinators almost everywhere in the world and forage at an impressive diversity of flowers. There are over 20,000 species of bee, but accurate information about their distribution and status in the world is very meagre. However, a recent study—"*Global patterns and drivers of bee distribution*" published in the journal *Current Biology*—has created a map of bee diversity which combines information from regional checklists with almost 6 million additional records from around the world. Compared with mammals and birds, bees are exceptional in that more species are found in dry, temperate areas than in the wet, tropical, forested environments that occur close to the equator. Hotspots for bee diversity include parts of the USA, Africa and the Middle East.

Bees have to visit flowers because, unlike their close relatives the wasps, they are vegetarian and feed both themselves and their larvae on pollen (a rich source of nutrients, including protein, fat and vitamins) as well as nectar (mostly converted to and stored as honey). They visit a prodigious number of flowers in order to gather sufficient resources to store in their nests. As a result, most bees are obliged to forage over relatively large areas, potentially making them important long-distance pollinators. Some tropical bees are known to be trapliners that fly at least 15–20 km every day, visiting widely dispersed but rewarding plants that open a few new flowers every day, often for several weeks or even months. They are key pollinators of flowers that are widely scattered but provide plentiful rewards.

Most of the flowers favoured by bees have a sweet scent but are often very different in other respects. Bees forage at flowers of almost any colour but tend to prefer yellow, blue and ultra-violet and often ignore red flowers because they are unable to 'see' red colours. The ability of bees to distinguish many different colours no doubt helps the flower constancy that they show when they repeatedly visit the most rewarding flowers—behaviour that is clearly benefits the flowers by helping to ensure that they are cross-pollinated.

Bees are influenced by the structure of flowers, as well as their colour, because their need for a secure place on which to land is particularly important. Most small bees prefer flattish or open saucer-shaped flowers, such as dandelions, buttercups, crane's-bills and brambles, all flowers that have an easily accessible, stable structure on which insects can land. Other flowers, such as hollyhocks, foxgloves, bindweeds and bellflowers, have a large, open entrance, spacious enough for most bees to alight safely in the mouth of the flower. Some bumblebees show a liking for bilaterally symmetrical flowers, such as Monk's-hood *Aconitum napellus*, snapdragons and claries—species with a robust lip that provides a secure place to land. The structure of many such flowers also ensures that visiting insects have to land in a specific position in order to get at the flower's nectaries or collect pollen. Getting into this position ensures that pollen is transferred accurately from the flower to the pollinator. Toadflaxes, broomrapes, and many legumes are other good examples of bilaterally symmetrical flowers that attract bees.

When it comes to collecting pollen, rather than nectar, bees are often very different in their flower preferences. The majority of species forage for pollen at a wide variety of different flowers. Others are more specialised and collect pollen from only a limited range of closely related flowers, often in a single genus. A few are even more specialised and almost invariably forage for pollen at just a single flower species. But note, while the more specialised bees are fussy about pollen, they readily visit other flowers for nectar. Among British bees, the Large Scabious Mining Bee *Andrena hattorfiana* and Small Scabious Mining Bee *Andrena marginata* collect pollen only at scabious species (Dipsacaceae); and the Small Scissor Bee *Chelostoma campanularum* and Gold-tailed Melitta *Melitta haemorrhoidalis* use only bellflowers *Campanula* spp. and close relatives. Species that collect pollen from only a single flower species include the Bryony Mining Bee *Andrena florea*, Sea Aster Bee *Colletes halophilus* and Ivy Bee *Colletes hederae* that visit and collect pollen only at White Bryony *Bryonia dioica*, Sea Aster *Tripolium pannonicum* and Ivy *Hedera helix* respectively. Such specialised bees are completely dependent on their foraging activity coinciding with the flowering seasonality of their host flower. Their survival depends on

Large Scabious Mining Bee *Andrena hattorfiana* at Field Scabious *Knautia arvensis* [PC]

Small Scabious Mining Bee *Andrena marginata* at Small Scabious *Scabiosa columbaria* [PC]

Small Scissor Bees *Chelostoma campanularum* at Harebell *Campanula rotundifolia* [PC]

Gold-tailed Melitta *Melitta haemorrhoidalis* at Clustered Bellflower *Campanula glomerata* [PC]

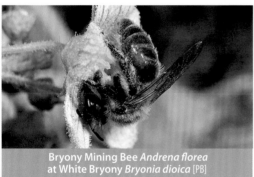

Bryony Mining Bee *Andrena florea* at White Bryony *Bryonia dioica* [PB]

Sea Aster Bee *Colletes halophilus* at Sea Aster *Tripolium pannonicum* [PC]

Ivy Bee *Colletes hederae* at Ivy *Hedera helix* [PC]

their chosen plant species having a long flowering season and being predictable in its seasonality year after year.

Bees also come in different sizes. The 275 species of British bees vary from as small as 6 mm long in some solitary bees, such as the Small Scissor Bee, to as large as 25 mm in the Violet Carpenter Bee *Xylocopa violacea*. Bees also have tongues that are very variable in their length. The majority of British species have a tongue that is only 0.5–5 mm in length and can forage for nectar only at flowers in which nectar is easily reachable. A few species, notably the Garden Bumblebee *Bombus hortorum*, Common Carder Bumblebee *Bombus pascuorum* and Hairy-footed Flower Bee *Anthophora plumipes*, have longer tongues (nearly 20 mm in length in the Hairy-footed Flower Bee) enabling them to get at nectar hidden deep in flowers such as columbines *Aquilegia* spp., Wallflowers *Erysimum cheiri*, Primroses *Primula vulgaris*, lungworts *Pulmonaria* spp. honeysuckles *Lonicera* spp., claries and sages *Salvia* spp. and toadflaxes *Linaria* spp. Elsewhere in the world, a few tropical orchid bees have a tongue more than 40 mm long and in some it is more than twice their body length.

Orchid Bee *Euglossa* sp. with very long tongue, Ecuador

Female bees are well adapted to collect and carry pollen back to their nest, where it is used to feed their larvae. Males do not collect pollen; nor do either sex of the many parasitic cuckoo bees (in many genera, including *Bombus*, *Coelioxys*, *Nomada* and others) whose larvae are cared for and fed by their hosts. Compared with most other insects, many bees are rather hairy, or positively furry in bumblebees. As they fly, hairy bees generate an electrostatic charge and, when they visit a flower, the charge attracts copious pollen onto their body and helps it remain in place. Later, female bees collect the pollen by grooming themselves and most stuff it into a storage organ, or pollen-brush, composed of modified hairs. Yellow-face bees *Hylaeus* spp. are exceptions that mix pollen with nectar and carry the mixture in their crop. Different bees often have their pollen brush on different parts of their body. Solitary bees in the subfamily Megachilinae, such as leaf-cutter bees *Megachile* spp. and mason bees *Osmia* spp., have a pollen-brush on the underside of their abdomen and scrape pollen onto it using their legs. Solitary bees in other genera, such as *Andrena*, *Lasioglossum* and *Halictus*, have a pollen brush on their hind legs, constructed using modified hairs. Honey Bees *Apis mellifera* and bumblebees *Bombus* spp. also transport pollen on their hind legs, packed in more elaborate pollen baskets.

Bumblebees are our most familiar and recognisable bees. Queen bumblebees are the only members of a colony to survive the winter and live on to found a new colony. Because she emerges early in spring, a queen first has to find enough flowers to give her a good enough meal to allow her to find a suitable nest site and begin egg-laying. The queen has to feed her first brood by herself without help. This results in a brood of tiny worker bees that take over the care of all subsequent broods later in the year, allowing the queen to concentrate on laying eggs.

Garden Bumblebee *Bombus hortorum*
at 'Hot-lips' *Salvia* [PC]

Hairy-footed Flower Bee *Anthophora plumipes*
at lungwort *Pulmonaria* [PC]

Large White-face Bee *Hylaeus signatus*
at Wild Mignonette *Reseda lutea* [PC]

Orange-vented Mason Bee *Osmia leaiana*
at crane's-bill *Geranium*

White-zoned Furrow Bee *Lasioglossum
leucozonium* at Cat's-ear *Hypochaeris radicata* [PC]

Honey Bee *Apis mellifera* at Common Rock-rose
Helianthemum nummularium

Recent research has revealed that queen bumblebees sometimes have trouble finding enough pollen early in the year. If pollen is scarce, it has now been established that bumblebees deliberately nibble holes in leaves, behaviour that results in plants flowering as many as 30 days earlier than they would have done. However, when experimenters deliberately damaged leaves themselves, the results were not fully replicated, suggesting that the bees must do something else to stimulate early flowering. An intriguing suggestion "*is that bee saliva might contain chemicals that prompt flowering—similar to chemicals in the saliva of plant-eating insects that prompt plant defence responses.*" Investigations continue.

Elsewhere in the world, there are bees with unusual modifications for collecting unusual rewards. In South Africa, for example, there are endemic bees *Rediviva* with long, hairy, front legs, adapted to collect oil when inserted into the long floral spurs of oil flowers *Diascia*. And in the Neotropics, male orchid bees have cushion-like or sponge-like pads of modified hairs on their legs, known as 'velvet patches', on which they accumulate the orchid perfumes that are irresistibly attractive to females.

Finally, it must be mentioned that Honey Bees are the only bee that have a 'dance language'. Successful scout bees dance after returning to the hive to pass on information to other hive members concerning any good flower resources that they have located. The performance tells hive members the direction and distance to the flowers and how abundant they are. Any scent remaining on the dancer identifies the kinds of flowers visited.

Pollination by wasps

Wasps are extremely diverse. Species with stinging ovipositors include the familiar social wasps (Vespidae), most of them social wasps but also including the solitary potter wasps (subfamily Eumeninae), all of the latter feeding their larvae by provisioning their nests with paralysed spiders, beetle larvae or caterpillars. Other solitary wasps (e.g. Pompilidae and Sphecidae) use their stinger-ovipositors to lay their eggs in paralysed insects or spiders that are then stored in their nests. Wasps also include numerous species with non-stinging ovipositors (e.g. in the families Ichneumonidae, Braconidae, Chalcidae, etc.) that are parasitic on other insects. Minute chalcid wasps, which lay their eggs in the eggs or larvae of other insects, are often beneficial, some species being used for biological control of agricultural pests. Parasitic wasps in the related family Trichogrammatidae are even more minute. Most species are less than 1 mm in length but are among the most important biological control agents of insect pests, particularly Lepidoptera. Many parasitise insect eggs and as many as 60 adult wasps have been recorded emerging from a single butterfly egg!

Most wasps differ from bees by being predatory or parasitic, for the most part feeding both themselves and their larvae on invertebrate prey (including many garden pests). Nevertheless, many adult wasps are frequent visitors to flowers in search of nectar as a source of supplementary energy. The genera *Vespula* and *Dolichovespula*, each of which include several species that resemble 'typical wasps', are excellent examples of wasps that consume liquid foods—nectar, honeydew, sap and body fluids of prey. In fact, it is now known that these wasps also deliver nectar and other liquid food to their larvae, not just animal food. In hot, dry weather, nectar and other sugary fluids are also valuable for their water content.

The majority of wasps have a short tongue and are only infrequent pollinators. Small wasps tend to visit the same small flowers that attract a multitude of diminutive flies, beetles and other unspecialised insects. Popular flowers include umbellifers, umbellifer-like species such as Yarrow *Achillea millefolium*, Elder *Sambucus nigra*, Blackthorn *Prunus spinosa* and Hawthorn *Crataegus monogyna* and numerous small-flowered composites. Compared with bees, most wasps are relatively hairless and too hard, smooth and shiny to trap much powdery pollen on their body. As a result, they are less effective than hairy bees as pollinators of conventional flowers. There are, however, a few British flowers that specialise in attracting vespid wasps. They tend to be dull reddish in colour and rather globular in shape. In appearance, figworts

Scrophularia spp. are classic examples of wasp-flowers though, in our experience, they usually attract many more visits from bumblebees than they do from wasps. Many orchids also make use of wasps as pollinators. The smooth bodies of wasps present no problems for orchids because their pollinaria are compact packages that are securely glued to their pollinators. More bizarre are the numerous European Bee Orchids *Ophrys apifera* that are pollinated by wasps (and some by bees) by pseudocopulation—a deceitful stratagem on the part of the orchids, that makes use of both visual and pheromone mimicry (see p. 54). Pseudocopulation is a deceit in which the orchids mimic female wasps or other pollinators and are pollinated when appropriate males attempt to mate with them. The Fly Orchid *Ophrys insectifera*, pollinated by solitary wasps, including the Two-girdled Digger Wasp *Argogorytes mystaceus*, is the best British example but there are many other species of bee orchids in southern Europe, North Africa and the Middle East that are pollinated by wasps, or sometimes bees. Other stratagems, both similar and different, are found in numerous other orchids in Australia and Central and South America.

Common Figwort *Scrophularia nodosa*, a classic example of a wasp-flower [PC]

Wasps also include the so-called sawflies—wasps that lack a 'wasp-waist'. Sawflies get their name from the females' saw-like ovipositor which they use to make incisions in the plants where they lay their eggs. There are about 500 species in the UK. Many sawflies are regular visitors to flowers to drink nectar but also feed on petals, stamens and pollen. Most sawflies have short mouthparts so they are more or less restricted to flowers with easily accessible nectaries. They are often seen on umbellifers and sometimes act as pollinators.

Tree Wasp *Dolichovespula sylvestris* with Broad-leaved Helleborine *Epipactis helleborine* pollinia [NO]

Turnip Sawfly *Athalia rosae* at Wild Carrot *Daucus carota*

Pollination by flies

Flies are thought to have been among the most important pollinators of the first flowering plants, presumably lured by pollen or the moisturising secretions of stigmas. Flies have since evolved to become extremely varied and numerous, with over 7,000 species in the UK and probably as many as a million worldwide (most of them still unclassified). Almost all flies have sucking mouthparts and can be divided into two main types. First, there are the biting flies, including mosquitoes, horseflies, sand flies and black flies, with mouthparts designed to pierce animal tissue and suck blood, though some supplement their diet with nectar and are occasional pollinators. For example, in the UK predatory flies, such as dance flies and the Yellow Dungfly, often hunt prey in the vicinity of flowers but also land on the flowers to feed on nectar and pollen.

The second and bigger group of flies includes hoverflies, flesh flies, blowflies, houseflies and many others, all with mouthparts that suck or mop up exposed fluids, such as nectar or fluids exuding from excrement or carcasses. Some can also ingest small, solid particles, including grains of pollen, provided they are suspended in saliva or nectar. Most of the flies in this second group are short-tongued, opportunistic visitors to unspecialised flowers (e.g. umbellifers, Yarrow and brambles) but others are key pollinators in cold, high-altitude or high-latitude habitats where bees are rare or nonexistent. Flies are then especially important because they often remain active on days that are too cold, wet or windy for bees. On the negative side, many flies are too small, and move only short distances between flowers, to be much use as cross-pollinators. Incidentally, the taste organs of flies occur in their legs as well as their mouthparts. The legs of blowflies are known to be up to 200 times more sensitive to sugar than the human tongue.

Hoverflies are very important flower visitors and usually thought to be behind only bees in their worldwide importance as pollinators. In fact, hoverflies, are known to visit at least 72% of global food crops as well as wildflowers, providing pollination and other services estimated to be worth around US$300 billion per year. It is also significant that several migratory species, that migrate to and from the UK from the Continent, regularly appear in enormous numbers. Unlike many other insect pollinators, hoverflies do not yet appear to be in serious decline.

The 273 species of British hoverflies are conveniently divided into three groups—the first includes species that feed mainly on pollen; the second with species that feed mainly on nectar; and a third with species that regularly feed on both pollen and nectar. Specialised pollen feeders, such as the Marmalade Hoverfly *Episyrphus balteatus* and species of *Eupeodes*, *Melanostoma* and *Syrphus*, have a short, thick tongue, only 2–4 mm long, and mostly visit yellow or white flowers with easily accessible nectar and pollen. The abundant Marmalade Hoverfly is one of very few hoverflies capable of crushing pollen grains as it feeds on them. Hoverflies that specialise on nectar, such as the snout hoverflies *Rhingia* spp. and the Hornet Hoverfly *Volucella zonaria*, have an elongated, slender tongue, 5–12 mm long, appropriate for probing and sucking nectar out of longish, floral tubes. The long tongue of these flies is conspicuous only when feeding. At other times it is folded into a space under the fly's head. Snout hoverflies easily reach the nectar of such tubular flowers as Red Campion *Silene dioica*, Primrose, Wallflower, Bugle *Ajuga reptans* and Bluebell *Hyacinthoides non-scripta*.

Hoverflies are sometimes said to be 'clever' flower visitors. Like bees, they too have the ability to use landmarks to navigate and to forage efficiently at diverse flowers, regardless of how they are constructed. For example, when feeding at a large daisy with many petals, at least some hoverflies are aware when they have completed a circuit around the flower and promptly depart.

Recent studies on hoverflies, including migratory species, clearly demonstrate the ecological importance of hoverflies, particularly two very abundant migrants—the Marmalade Hoverfly and the Vagrant Hoverfly *Eupeodes corollae*, both of which provide important pollination and pest control services. Using insect-monitoring radars, the researchers were able to show that up to 4 billion hoverflies (equivalent to 80 tons of biomass) migrate at high altitudes in and out of the UK from Europe every year and contribute enormously to the pollination of both

A dance fly *Empis tessellata* feeding on nectar at Hawthorn *Crataegus monogyna* flowers

Glass-winged Syrphus *Syrphus vitripennis* feeding on pollen at a rock-rose *Cistus* sp.

Common Snout Hoverfly *Rhingia campestris* at Honesty *Lunaria annua*

Stripe-winged Dronefly *Eristalis horticola* at Firethorn *Pyracantha coccinea*

Marmalade Hoverfly *Episyrphus balteatus* at Wild Parsnip *Pastinaca sativa* [PC]

Hornet Hoverfly *Volucella zonaria* at Hemp-agrimony *Eupatorium cannabinum* [PC]

wildflowers and flowering crops. In addition, it is calculated that the hoverflies' voracious larvae contribute to the control of crop pests by consuming 6 trillion or more aphids. This huge number of migratory, flower-visiting hoverflies means that during the summer they rival managed Honey Bees in abundance, and probably as pollinators. There are up to 4 billion migrant hoverflies in just southern Britain during May to September compared with about 5 billion managed Honey Bees in the whole of the UK.

Unlike hoverflies, with their fold-away tongues, bee-flies *Bombylius* spp. have a long, rigid tongue (proboscis), 10 mm or more long, that enables them to feed at flowers with a very long nectar tube. Bee-flies usually hover when feeding but, even while hovering, often fine tune the aim of their long, probing tongue by steadying themselves with their legs. Bee-flies are as furry as bumblebees and pollen is easily trapped amongst their hairs. In the UK, Dark-edged Bee-flies are among the earliest pollinators to emerge in spring. They are frequent visitors to blue, orange, or yellow flowers, favourites including Aubretia *Aubrieta deltoidea*, Honesty *Lunaria annua*, forget-me-nots *Myosotis* spp. and grape-hyacinths *Muscari* spp., but occasionally visit the blossom of plum, cherry, other fruit trees or even Blackthorn. In South Africa, there are a number of even more specialised flies, notably tangle-veined flies, with even longer, rigid tongues

Dark-edged Bee-fly *Bombylius major*
at Aubretia *Aubrietia*

Flesh Fly *Sarcophaga carnaria*
at Canadian Goldenrod *Solidago canadensis*

Common Greenbottle *Lucilia caesar*
at a yellow composite (Asteraceae)

Blue-winged Tachinid *Phasia hemiptera*
at Wild Carrot *Daucus carota* [PC]

than bee-flies—up to 70 mm long and about four times the length of their body. Their tongue enables them to reach the nectar at the bottom of the extra-long floral tubes of painted petal irises *Lapeirousia*.

Of the many other British flies, the most common visitors to flowers are members of several families, including blowflies (Calliphoridae), flesh flies (Sarcophagidae) and tachinid flies (Tachinidae), that superficially resemble large, colourful house flies. All feed at flowers with easily accessible nectar and all are effective pollinators at times. Familiar blowflies include bluebottles *Calliphora* spp. and greenbottles *Lucilia* spp. The type specimen of the bluebottle goes by the evocative name *Calliphora vomitaria*. Blowflies lay their eggs on rotting carcasses and their larvae are often used in forensic entomology to determine the time of death of murder victims. Blowfly larvae are also used for 'maggot therapy' to clean dead flesh in open wounds. I was once in hospital and scheduled for 'maggot therapy'. It would have been a excellent natural history experience, so it was most disappointing when the maggots failed to arrive in time. Flesh flies are typically black and grey with stripes on their thorax and a checker-board pattern on their abdomen. They are common visitors to flowers but also feed on liquid seeping from rotting carcasses or excrement. They differ from most flies, including blowflies, in being ovoviviparous, depositing newly hatched maggots (not eggs) on carrion, excrement or even on open wounds of injured mammals. Tachinid flies are noticeably bristly and often colourful. The adults feed on honeydew and nectar and are common visitors to many different flowers. Almost all tachinid flies have parasitic larvae. The flies lay their eggs in or on the larvae of butterflies, moths, beetles, sawflies and other groups. The host is always killed sooner or later.

Numerous flies are important pollinators elsewhere in the world and it may be of interest to chocolate lovers to know that tiny midges are indispensable to the chocolate industry. Even though cocoa trees *Theobroma cacao* have been cultivated in Central America for over a thousand years, first of all by the chocolate-loving Aztecs, their flowers still need to be pollinated by midges.

Finally, there are numerous aroids, mostly tropical species, that use mimicry and deceit to attract flies as pollinators. Flies enter and become trapped in inflorescences that mimic sites that are suitable for laying eggs in their both appearance and smell. The aroid, Lords-and-ladies *Arum maculatum*, is a common British example. It has an inflorescence, or spadix, surrounded by a leaf-like spathe. The spadix is covered by a band of female flowers at the base and male flowers higher up. The stigma of female flowers is receptive only before pollen is shed by the male flowers, ensuring that self-pollination is avoided. The inflorescence lasts for most of two days. On the first afternoon, the spadix heats up to about 15° above the ambient temperature, releasing a foul odour, and attracts owl midges that become trapped beneath a ring of bristles. Some of the midges arrive carrying pollen from an inflorescence visited earlier and pollinate the receptive female flowers. They remain trapped long enough to become dusted with pollen from newly ripened stamens of male flowers, after which they make their escape past the now-wilted bristles and perhaps visit and cross-pollinate the female flowers in another newly opened inflorescence.

Pollination by beetles

Beetles, together with flies, are thought to have been the original pollinators of the first ancient flowers. They certainly lived alongside the earliest flowers because fossils prove that beetles arrived on the scene more than 270 million years ago, long before the first flowering plants (angiosperms) which are thought to have first emerged around 130 million years ago. The mouthparts of beetles are designed for chewing, so beetles tend to be messy, destructive feeders that routinely munch on stigmas, stamens and other flower parts, as well as consuming pollen and nectar. Because of the need to minimise damage by beetles, the ovaries of beetle-flowers are usually enclosed and protected by robust, thickened carpels.

There are plenty of specialised beetle pollinators in tropical areas of the world but none of the 3,700 species found in temperate Britain are particularly important as pollinators. There

Black-and-yellow Longhorn Beetle *Rutpela maculata* at Rough Chervil *Chaerophyllum temulum*

Swollen-thighed Beetle *Oedemera nobilis* at Common Knapweed *Centaurea nigra*

are, however, numerous small beetles that forage in unspecialised flowers together with a multitude of flies, wasps and other insects. Common British examples include the Garden Chafer *Phyllopertha horticola*, Common Red Soldier Beetle *Rhagonycha fulva*, Swollen-thighed Beetle *Oedemera nobilis* and Black-and-yellow Longhorn Beetle *Rutpela maculata*. In common with larger beetles, they are messy feeders and just as likely to damage flowers as pollinate them. Most beetles do not fly about as actively as most other insect pollinators and seldom move pollen very far. The tiny pollen beetles that are so abundant in temperate regions tend to stay for several hours or even days in the same flower and are probably insignificant as pollinators.

Elsewhere in the world, there are two particularly important groups of specialised beetle-pollinated flowers known as 'chamber blossoms' and 'painted-bowls'. Chamber blossoms are mainly found in tropical rainforests and familiar examples include *Magnolia* species and numerous aroids similar to Lords-and-ladies (mentioned above). The other group of beetle-flowers are appropriately called painted-bowls. As suggested by their name, the flowers concerned are bowl-shaped and colourful. Painted-bowl flowers are common and widespread in hot, dry regions. They are abundant, for instance, in Namaqualand (in South Africa and adjoining Namibia), where many of the colourful flowers that bloom in glorious profusion following the winter rains, are pollinated by scarab monkey beetles.

Pollination by butterflies

There are just 59 resident species of butterflies that breed in the UK. Adult butterflies feed on liquids, including nectar, honeydew, tree sap, juice from over-ripe fruits and fluids exuding from excrement and carrion. The liquids are sucked up through the long proboscis which is coiled when not in use. The proboscis of feeding butterflies has a characteristic, adjustable 'knee-bend' that facilitates its entry into slender, tubular flowers. Most British butterflies feed mainly on nectar, though holly blues and several woodland species regularly supplement their diet by feeding on tree sap and honeydew secreted by aphids. Peacocks *Aglais io* and Red Admirals *Vanessa atalanta* also gather at over-ripe plums or rotting apples, while White Admirals *Limenitis camilla* and Purple Emperors *Apatura iris* are attracted to dung and animal carcasses, including road kills.

Butterflies are sun-loving insects that like to sit and bask in warm sunshine whilst sipping nectar. Butterflies visit flowers that are diurnal with a sweet smell. And butterfly-flowers must have a convenient platform on which to land the sort of place provided by inflorescences with densely packed tiny flowers or individual flowers shaped like a shallow saucer. Favourite plants in the UK are mainly perennials and include Buddleia, Hemp-agrimony *Eupatorium cannabinum*, Honesty, crane's-bills *Geranium* spp., lavender *Lavandula* spp., Wild Marjoram *Origanum vulgare*, Red Valerian, brambles *Rubus* spp., scabiouses (Dipsacaceae) and coneflowers (Asteraceae). Most flowers preferred by butterflies produce average amounts of nectar, often secreted within a long, thin spur that matches the long, thin tongue of butterflies. In temperate zones, the flowers preferred by butterflies tend to be purple, lilac, pink or yellow. However, many butterflies have the ability to see a wider range of wavelengths, from ultra-violet to red, than most other insects. Red butterfly flowers are common in the tropics where many are made even more conspicuous and attractive to butterflies by being surrounded by red ancillary structures, including calyces, bracts, or leaves. Poinsettia and Wild Poinsettia *Warzewiczia coccinea* are excellent examples.

Painted Lady *Vanessa cardui* at Buddleia *Buddleja davidii*

Pollination by moths

There are two well-defined types of flowers that are pollinated by moths. Members of both groups have flowers that open at night and are white or at least pale, making them visible in the low light conditions of dusk, dawn, or in moonlight. They also have a sweet fragrance that is a compelling allure for moths and usually much stronger than the scent of butterfly flowers. The flowers in one of these groups do not provide a landing platform and usually have long, nectar-spurs that secrete copious, rather dilute nectar. They are intended to attract hawk-moths (Sphingidae) that habitually hover to feed. The other group of flowers typically provide a suitable place to land and have nectaries that are readily accessible. Most are visited and pollinated by moths that lack the ability to hover and have to settle on a flower to feed.

In many respects, the hawk-moths in the first group are like hummingbirds. They share the same superb hovering skills and have a tongue that more than matches the long bill and tongue of hummingbirds. Many tropical hawk-moths are known to move several kilometres at night in search of flowers, making them excellent long-distance pollinators. There are numerous hawk-moths with a tongue 40–100 mm long and a few with an extraordinary tongue, several times that length. Charles Darwin famously predicted the existence of a hawk-moth that must

Silver Y *Autographa gamma* feeding at Lavender *Lavandula*

Six-spot Burnet *Zygaena filipendulae* at Field Scabious *Knautia arvensis* [PC]

Hummingbird Hawk-moth *Macroglossum stellatarum* at Lavender *Lavandula*

Broad-bordered Bee Hawk-moth *Hemaris fuciformis* at Lavender *Lavandula* [PC]

have a tongue long enough to pollinate a recently described Madagascan orchid with a nectar spur measuring over 400 mm in length. Darwin was eventually proved right when a previously unknown hawk-moth with an exceptionally long tongue was collected and described.

Many of the moths in the second group are short-lived, lack mouthparts and do not feed at all. But others, most of them in the families Noctuidae and Geometridae, are flower visitors but have to land on a flower to feed. They have a shorter tongue than hawk-moths and seek out strongly scented flowers that have easily accessible nectar. Good British flowers include pinks *Dianthus* spp., the paler varieties of Phlox *Phlox paniculata*, and White Campions *Silene latifolia*. The visiting moths are crepuscular but mostly fly at dusk rather than dawn, when temperatures are still warm after a sunny day. In the UK, for example, the European Silver Y moth *Autographa gamma* is an abundant visitor and comparable noctuid moths are found wherever suitable flowers exist. Silver Y moths are migratory, often arriving in the UK from the continent in enormous numbers. They typically feed at dusk but, during an invasion, can even be seen feeding throughout the sunniest, hottest afternoons. Popular flowers then include Honeysuckle, lavender, knapweeds *Centaurea* spp. and Red Valerian.

Compared with bees and hoverflies, nocturnal macromoths tend to be neglected in pollination research, at least in temperate regions. However, a recent paper shows that they "*have an important but overlooked ecological role.*" According to lead author, Dr Richard Walton, they complement the activity of diurnal pollinators. "*While bumblebees and honeybees are known to be super pollinators they also preferentially target the most prolific nectar and pollen sources. Moths may appear to be less effective pollinators by comparison, but their high diversity and abundance may make them critical to pollination in ways that we still need to understand.*" The study showed that pollen loads were most often found on the ventral thorax of moths, not their proboscis, and the frequency with which pollen loads were found suggests that nocturnal macromoths provide vital pollination services for several wild plant families in agricultural landscapes. Given that moth populations have suffered severe declines since the 1970s (see p. 17), mainly due to changes in land use and the overuse of pesticides, it is clearly important to include nocturnal macromoths in future management and conservation planning.

Of course, not all moths are nocturnal. Day-flying moths in the UK include the Hummingbird Hawk-moth *Macroglossum stellatarum*, Narrow-bordered Bee Hawk-moth *Hemaris tityus* and Broad-bordered Bee Hawk-moth *Hemaris fuciformis*, as well as several species with conspicuous warning colours, such as the Cinnabar moth *Tyria jacobaeae* and several burnet moths *Zygaena* spp. Most day-flying moths behave more like butterflies than typical moths and tend to forage at the same flowers. Favourites of the Hummingbird Hawk-moth, for example, include Buddleia, Lavender and Red Valerian, all of which are very popular with butterflies.

Pollination by wind

Wind pollination can be very effective if and when plants of the same species are growing in close proximity and flowering at the same time. Wind pollination is particularly widespread and effective in plants growing together in relatively open habitats, such as the canopy of temperate woodland or grassland.

Compared with insect-pollinated flowers, wind-pollinated species are usually simpler in form and colour. They do not need to attract pollinators and so have no need to be colourful, to emit odours or offer rewards. As a result, their floral parts, especially their petals, are almost always much smaller than those of insect-pollinated flowers or often missing. Indeed, the flowers of grasses and lots of temperate trees, including oaks, elms, birches, ashes and Hazel, are so inconspicuous that they barely resemble normal flowers at all. On the other hand, the anthers of wind-pollinated flowers tend be relatively large with their pollen exposed to wind or, as in numerous grasses, with their anthers dangling on slender filaments that tremble in even the gentlest puff of air, releasing clouds of pollen. Wind-dispersed pollen is typically dust-like and so light that it is readily carried a long way by even a gentle breeze. Also, the stigmas of wind-pollinated flowers tend to be exposed and feathery, and often coated with a sticky secretion. They are beautifully adapted to intercept and capture pollen carried by the wind. It has also been established that electrostatic attraction may also play a part in capturing pollen.

In temperate woodland, trees generally flower early in the year, in March or April, before they leaf out, so that the movement of pollen carried on the wind is unimpeded by foliage. The time of day when pollen is released is also important. The flowers of temperate trees typically release their pollen in the afternoon, the time of day when the air tends to be at its warmest and driest.

The stigma of wind-pollinated flowers is usually feathery. Nevertheless, it is tiny, often not much more than a square millimetre in area. Hence, to be successful, wind pollination requires vast amounts of pollen to be released to drift on air currents. Indeed, the amount of pollen released is often hard to believe—a Hazel catkin releases up to four million pollen grains and some grasses release as many as ten million grains per flower-head. This is, of course, one reason why so many people are badly affected by hay fever in spring and summer.

Pollination by water

There are very few plants that rely on water pollination because the viability of pollen seldom survives immersion in water. That is why the majority of plants that grow in water have flowers on stems that emerge from the water and bloom where they can be pollinated just like flowers that grow on land—by wind or insects. Well-known aquatic plants that are pollinated by insects include Flowering-rush *Butomus umbellatum*, water-crowfoots and spearworts *Ranunculus* spp., Bogbean *Menyanthes trifoliata*, Arrowhead *Sagittaria sagittifolia*, Yellow Iris *Iris pseudacorus* and water-lilies *Nuphar lutea* and *Nymphaea alba*. The only common British plants with flowers that are pollinated in water are eelgrasses *Zostera* spp. and related marine plants.

CHAPTER 14
Mimicry and deceit in flowers

In spite of the obvious mutual benefits, the interplay between plants and their pollinators is not a real partnership. The relationship is entirely self-serving, without even the slightest trace of genuine collaboration. Plants and their pollinators take advantage of each other if they can get away with it or else interact with the minimum expenditure of resources. Plants cheat because it allows them to avoid or reduce the cost of producing nectar, pollen or other nutritious rewards. And, as already mentioned, pollinators often try to maximise the value of nectar, pollen or other rewards by stealing them with the minimum expenditure of effort.

Deception by flowers is most frequent in sparsely distributed species that do not profit from the flower-constancy behaviour shown by many bees and some other insect pollinators. The most prolific cheaters are the orchids. There are well over 20,000 species of orchids in the world and very nearly a half have no reward to offer to pollinators, relying instead on deceit. Numerous orchids are just generalised flower mimics that emit flower-like signals that attract at least a few pollinators. They mostly rely on inexperienced pollinating insects, such as those newly emerged from pupae, treating them as flowers that really do offer a reward of nectar or pollen. And there are others that display an array of dishonest stratagems, including both visual and pheromone mimicry. The most extreme examples involve pseudocopulation, a deceit in which orchids mimic key features of female bees, wasps or other pollinators and are pollinated when appropriate males are attracted and attempt to mate with them. Pseudocopulation occurs in many European bee orchids *Ophrys* spp. as well as numerous orchids in Australia and Central and South America. It must be emphasised that plants that cheat are most likely to succeed and benefit when there is an abundance of newly emerged, inexperienced insect visitors on hand, just waiting to be fooled. Flowering at the right time is important though it only takes a few visitors to achieve the cross-pollination that promotes genetic variability.

The deceitful mimetic structures used by flowers also include pseudoflowers, pseudopollen, and pseudonectaries—structures that attract pollinators to a flower (or inflorescence) by appearing to advertise or offer rewards. Guelder-rose *Viburnum opulus* and

Pseudocopulation: a Two-girdled Digger Wasp *Argogorytes mystaceus* attempting to mate with and pollinate a Fly Orchid *Ophrys insectifera* [NO]

Guelder-rose *Viburnum opulus* with pseudoflowers [PC]

A Monkeyflower *Erythranthe guttata* with hairy ridges on its lip – pseudopollen

Grass-of-Parnassus *Parnassia palustris* has five three-stalked staminodes (pseudonectaries), each tipped with a glistening drop-like protuberance [PC]

hydrangeas *Hydrangea* spp. are attractive plants, both of which have umbel-like inflorescences that are decorated around their margins with large, ornamental pseudoflowers. These eye-catching but sterile flowers offer no reward but certainly enhance the visual display and so help attract pollinators. Pseudopollen exists in several variations, including eye-catching staminodes that look like real anthers, and yellow markings that suggest the presence of anthers. Examples of pseudopollen can be seen in the yellow 'beard' of the Bearded Iris *Iris germanica* and the hairy ridges on the lip of the Monkeyflower *Erythranthe guttata*. Such vaguely pollen-like characters target the innate visual preferences of potential pollinators and may deceive them into 'taking a closer look'. Sometimes the deception works. Pseudonectaries are much less common than pseudopollen but Grass-of-Parnassus *Parnassia palustris* flowers provide an outstanding example. The flowers have five staminodes each of which has three stalks tipped with a glistening, drop-like protuberance. These apparent 'nectaries' are dry but so visually realistic and attractive that inexperienced visiting flies are often deceived.

PART 5
Flowers for pollinators

Wildflower meadow featuring Corn Marigolds *Glebionis segetum*

CHAPTER 15
Recommended flowers

In the following pages we have selected about 100 attractive flowers, mainly native or naturalised British species, most of them personal favourites that grow in our own garden (in Cley-next-the-sea on the North Norfolk coast) and that we have found to be good for pollinators. We have often given preference to plants that are also good for wildlife in other ways—for example, by having wildlife-edible fruits, foliage that is a preferred food of Lepidoptera larvae, or by providing nesting cover for birds. We have also included a few non-native garden flowers that are especially good for pollinators, most notably lavender *Lavandula* spp., Buddleia *Buddleja davidii*, Tansy-leaved Phacelia *Phacelia tanacitifolia* and a coneflower variety (Asteraceae). Several of our choices, including Bramble *Rubus* spp., Buddleia and Wall Cotoneaster *Cotoneaster horizontalis*, are potentially invasive. They are exceptional plants for pollinators and other wildlife but need to be well managed. Our selection also includes unpopular 'weeds', such as thistles and ragwort, and several plants that are poisonous, all of them outstanding for pollinators. All deserve to be encouraged, even if confined to a small, out of sight corner of the garden. Bear in mind that many of our most popular and familiar garden plants are also poisonous, including such favourites as Monk's-hood *Aconitum napellus*, Pasqueflower *Pulsatilla vulgaris*, Larkspur *Consolida ajacis*, Corncockle *Agrostemma githago*, Foxglove *Digitalis purpurea*, daffodils *Narcissus* spp. and crocuses *Crocus* spp.

Our aim is a selection of plants that will attract a dependable diversity of both small and large pollinators, including flies and small bees, not just bumblebees, butterflies and other large species. Umbellifers, for example, attract very few of the larger pollinators but are excellent for lacewings, ichneumon wasps, sawflies, small solitary bees, and numerous small flies and beetles. The umbellifer-like Yarrow *Achillea millefolium*, a member of the daisy family, is also quite good.

We have also made a point of including several plants that flower early in the year—in February and March—and so provide a valuable source of nectar for insects that emerge early, especially queen bumblebees and butterflies that have overwintered. The most useful early flowers in our garden include Goat Willow *Salix caprea*, Blackthorn *Prunus spinosa*, Rosemary *Salvia rosmarinus*, Dandelion *Taraxacum* spp., Green Alkanet *Pentaglottis sempervirens*, Flowering Currant *Ribes sanguineum* and three ornamental crucifers—Aubretia *Aubrieta deltoidea*, Honesty *Lunaria annua* and Wallflower *Erysimum cheiri*. There are numerous other early-flowering plants, notably snowdrops *Galanthus* spp., daffodils, crocuses, hellebores *Helleborus* spp. and Winter Aconite *Eranthis hyemalis*. They are pleasing to the eye but, given the vagaries of the weather in an era of climate change, often come into flower so early that they seldom (at least in our experience in recent years) attract many insects. Late flowering plants are also important. Some of the most useful, notably Ivy *Hedera helix*, Butterfly Stonecrop *Hylotelephium spectabile*, *Caryopteris* and Yarrow, continue flowering from August (or even earlier) until late October or even November. Our final selection includes only a few monocots. None of those that grow in our garden, including Bee Orchid *Ophrys apifera*, daffodils, Spring Crocus *Crocus vernus*, Bluebell, grape-hyacinths *Muscari* spp. and spiderworts *Tradescantia* spp., attract pollinators in large numbers.

In the following family and species accounts, the sequence and taxonomy of British native plants, naturalised plants and common garden escapes follows the *New Flora of the British Isles* (4th edition, 2019) by Clive Stace, the standard work for the British flora. However, Stace does not necessarily follow all the most recent changes and recommendations made by the Angiosperm Phylogeny Group IV (2016), based on phylogenetic studies using evidence from DNA sequencing. Here, we have followed Stace's taxonomy and nomenclature throughout but have also mentioned recent APG IV changes and recommendations that have not yet been followed by Stace.

DICOTYLEDONS

Papaveraceae (Poppy family)

According to recent research, the poppy family now includes the fumitories (formerly in the Fumariaceae) and *Pteridophyllum racemosum* (a species endemic to Japan, formerly in the Pteridophyllaceae. Worldwide, the new Papaveraceae classification includes about 40 genera and almost 1,000 species. This enlarged family is found almost worldwide, but is commonest in northern temperate regions and is poorly represented in the tropics. The members of the family are mostly herbaceous annuals or perennials though there are a few shrubs and small trees.

In the earlier version of the Papaveraceae, most of the typical 'poppies' were in just three genera—*Papaver* with about 80 species mostly in northern temperate regions; *Meconopsis* with 48 species and a centre of diversity in the Himalayas and adjacent China; and *Argemone* with 32 species native to the Americas. Numerous members of the family are valuable garden ornamentals and pharmaceutically important plants. Brightly coloured yellow, orange or white latex, containing powerful narcotic alkaloids, is characteristic of many species of *Papaver*. Species with medicinal value include the Opium Poppy, Corn Poppy and Yellow Horned-poppy (see species accounts below).

Here we are concerned only with a few 'poppies' that occur in the UK as natives or ancient introductions. The only native UK poppies are the Yellow Horned-poppy and Welsh Poppy. The ancient introductions include five red poppies and the Opium Poppy, all of them introductions that arrived in 'seed-corn' in the Bronze Age, around 4,000 BC, or perhaps even earlier. There are numerous ornamental garden poppies of which the most spectacular must be the huge blue Himalayan poppies *Meconopsis* spp.

Poppy flowers are radially symmetrical, shallow bowl-shaped and usually have two sepals and four petals. Poppies produce no nectar. They are pollen-only flowers and bumblebees are the principal pollinators. The fruit is a variously shaped capsule, the shape being a good aid to identification. The tiny seeds are dispersed from holes at the top of the 'pepper shaker-like' capsule when the plant is shaken by the wind.

Opium Poppy *Papaver somniferum*

The Opium Poppy is thought to be native to the eastern Mediterranean region but, because of ancient introductions and cultivation, it is now naturalised across much of Europe and Asia. It has probably been cultivated in Britain since the Bronze Age and is now quite common in and around towns, villages and on waste ground.

The Opium Poppy's fruit is a capsule that exudes a latex containing many pharmaceutically important alkaloids, including opium, codeine and morphine—the latter the source of heroine. The latex is collected by making incisions in immature capsules. In modern medicine, morphine is used as a pain killer, particularly in terminal cases or in patients enduring severe pain. Codeine is used in cough syrups as a mild sedative. It should be noted that plants growing in the British climate produce very little in the way of narcotic alkaloids.

Opium Poppy seeds are harvested and widely used in food products as a spicy condiment or seed decoration sprinkled on baked goods, including bread, pastries, muffins

Opium Poppy and Marmalade Hoverfly
Episyrphus balteatus

and bagels. It has been suggested that 'breadseed poppy' would be a more appropriate name than 'opium poppy' because all varieties produce edible seeds, whereas some varieties produce little or no opium. In 2016, world production of poppy seeds was over 92,000 tons with the Czech Republic, Turkey and Spain being the major producers. This is an extraordinary statistic considering that it takes between one and two million seeds to weigh a pound.

The Opium Poppy is quite often encountered on cultivated land or wasteland throughout much of the UK. It is an annual that grows to a metre or more in height. It flowers from about June until August. The large flowers, 10–18 cm across, sometimes much smaller, are most commonly lilac with a purple centre, but some varieties are white or red. They are very attractive to bumblebees and pollen-eating hoverflies.

Corn or Common Poppy *Papaver rhoeas*

The Corn Poppy and four other 'look-alike' red poppies are found in the UK. All five are ancient introductions that arrived in 'seed-corn' at least as long ago as the Bronze Age, perhaps earlier with Neolithic farmers around 4,000 BC. Poppies are agricultural weeds that thrive on disturbed ground, which is why the Corn Poppy bloomed so profusely all across the devastated land of the Flanders battlefields and came to symbolise the blood spilled in the war. This evocative image of the poppy is encapsulated in the poignant poem by Canadian military physician Colonel John McCrae:

Corn Poppies

> *In Flanders Fields the poppies blow,*
> *Between the crosses, row on row,*
> *That mark our place; and in the sky*
> *The larks, still bravely singing, fly*
> *Scarce heard amid the guns below.*

Like other poppies, the Corn Poppy contains various alkaloids though they are much milder in their effect than those of the Opium Poppy. In traditional folk medicine, the Corn Poppy was used to treat insomnia, coughs and miscellaneous aches and pains. And the petals were used to make a syrup to help children sleep.

The Corn Poppy's original native range is not known for certain but it is probably native to southern Europe and the Mediterranean region. It has been associated with the spread of agriculture for so long that it now has an enormous range in temperate areas of Europe, Asia, northern Africa and North America. In England the Corn Poppy is common below about 400 m, but scarcer and more local in Wales, Scotland and Ireland.

Intensive agriculture and the use of modern herbicides has caused many 'cornfield annuals', including the Corn Poppy, to decline drastically in the wild. However, corn poppies produce tens of thousands of seeds, many of which remain dormant in the soil for 80 years or more. It is this seed bank that results in the spectacular displays of poppies that sometimes appear nowadays on set-asides and field margins where they have long been absent.

The Corn Poppy is an annual plant that grows up to about 60 cm high. It flowers from June to August and can produce as many as 400 flowers in a good season. The showy flowers are 6–10 cm across with four vivid, red petals often with a black blotch on their base. The flowers last only a day and are capable of self-pollination. Poppies are pollen-only flowers that lack nectar. The abundant pollen attracts many bees and a few of the hoverflies that are adapted to eat pollen.

Welsh Poppy *Papaver cambricum*

Welsh Poppy

The Welsh Poppy is a perennial plant that is native to mountainous regions of western Europe, including the UK, southern France, the Pyrenees and the Iberian Peninsula. It is native to much of Wales and south-western England. However, the Welsh Poppy is a popular garden plant and frequent escape and it is now widely naturalised, particularly in northern England and Scotland.

Although it was originally named *Papaver cambricum* by Carl Linnaeus, it was later made the type species of the new poppy genus *Meconopsis* that later came to include the many species of 'blue poppies' that were discovered in the Himalayas and adjacent areas of China. However, recent phylogenetic research, using evidence from DNA sequencing, has shown that the Welsh Poppy is closer to species of *Papaver* than to the Himalayan species of *Meconopsis*. It is once again placed in the genus *Papaver*.

Nowadays, the Welsh Poppy is widely distributed and naturalised in the UK, particularly in the more hilly regions found in the north and west. It is found in damp woodland, rocky hillsides, shady gullies and cliff ledges. As a common garden escape, it is widespread and well-adapted to colonise urban niches, including shaded hedge banks, roadside verges and old walls.

The Welsh Poppy grows to about 60 cm tall with graceful foliage and golden yellow or orange flowers with four petals measuring about 4–6 cm across. Flowers last only one day but are produced in succession for several months from May until August. The flowers lack nectar but produce abundant pollen and are very attractive to bumblebees.

Yellow Horned-poppy *Glaucium flavum*

The Yellow Horned-poppy is native to temperate regions of Europe, North Africa and adjacent parts of western Asia. It is also found in Macaronesia, four archipelagos in the northern Atlantic—the Azores, Madeira, the Canary Islands and Cape Verde Islands. It has been introduced to North America, where it is listed as a noxious weed. In the UK, it is a coastal plant found as far north as the Wash on the east coast and the Solway on the west coast. It grows on

Yellow Horned-poppies, Norfolk, and flower with a female Swollen-thighed Beetle *Oedemera nobilis*

the seashore, mainly on shingle banks above the high tide line, and is never found inland (except in gardens). The Yellow Horned-poppy is protected under the Wildlife and Countryside Act 1981. It must not be picked in the wild.

The Yellow Horned-poppy exudes a foul-smelling, orange sap when cut or damaged. All parts of the plant are toxic and cause a range of problems if eaten, which are said to include brain damage and respiratory failure, sometimes resulting in death. In the past, oil extracted from the seeds was used to make soap and also burned in oil lamps.

The Yellow Horned-poppy can be a herbaceous biennial or perennial. It has rosettes of grey-green leaves which are protected from dehydration by a layer of water retaining wax. Its spreading, multi-branched stems reach heights of up to 90 cm. It blooms from June until September and its golden-yellow, four-petalled flowers are 60–90 mm across. The flowers are followed by slender, curved seedpods, up to 30 cm long—these 'horns', which give the species its name, are the longest seedpods of any UK plant.

The pollen-only flowers are very attractive to many bumblebee species and a few hoverflies. Although poisonous to humans, the seeds are eaten by finches and buntings, notably the Linnets, Twites and Snow Buntings that are winter visitors on the UK's coasts.

Ranunculaceae (Buttercup family)

The family Ranunculaceae includes about 2,500 species in 60 genera. It is one of the most cosmopolitan families though the majority of species are found in temperate regions of the Northern Hemisphere or at high altitudes in the tropics. There are very few species in lowland tropical rainforest. Almost all members of the family are herbaceous, the main exceptions being the woody climbers in the genus *Clematis*.

In the UK, including naturalised garden escapes, there are about 17 genera and over 40 species. Many are familiar wildflowers or cultivated garden flowers, including buttercups *Ranunculus* spp., Marsh-marigolds, anemones *Anemone* spp., hellebores *Helleborus* spp., Love-in-a-mist *Nigella damascena*, columbines, delphiniums and Monk's-hood. Most species have radially symmetrical flowers, though floral structure is nevertheless very variable, so much so that it is sometimes difficult to believe that all family members really belong together. The buttercups have a conventional structure with a normal calyx and five petals but in numerous other species, including Pasqueflowers, most anemones, Marsh-marigolds and Globeflowers *Trollius europeaus*, true petals are missing and it is the calyx and its sepals that are brightly coloured and 'flower-like'. Species of meadow-rue also lack petals though some have small petal-like sepals. However, their main attractive floral display is provided by numerous long, colourful stamens. The structure of columbines and love-in-a-mist is more complex and very unlike that of buttercups and anemones. The same is true of the bilaterally symmetrical members of the family, notably Monk's-hood and delphiniums.

Many members of the Ranunculaceae have very showy flowers and are cultivated as ornamental garden flowers. Pasqueflowers, anemones, hellebores, Love-in-a-mist, columbines, delphiniums and Monk's-hood are popular examples. Many members of the family are very poisonous. Nevertheless, some species are used as herbal medicines and in homeopathy (see species accounts).

Marsh-marigold or Kingcup *Caltha palustris*

The Marsh-marigold is a herbaceous perennial that is native to temperate regions of the Northern Hemisphere. In the UK, it is found virtually throughout, below about 1,000 m. It is larger, more thickset and more impressive than its buttercup relatives and a common plant of pond margins, fens, marshes, damp meadows and wet woodland. Marsh-marigold is an excellent plant to add to a wildlife-friendly pond, because it provides shelter for amphibians and early nectar for insects.

The generic name *Caltha* is derived from the Greek for 'goblet' and refers to Marsh-marigold's golden flowers being 'cups fit for kings'. Hence the alternative name "Kingcup".

Marsh-marigold

Marsh-marigolds grow up to about 60 cm tall and have dark, glossy green leaves. The flowers, 20–50 cm across, lack petals but their 5–8 glossy, golden-yellow sepals provide a welcome splash of colour in early spring. They bloom from late February or March until June and provide an early source of pollen, as well as nectar, for many insect visitors, including bees, hoverflies and smaller numbers of butterflies and beetles. Apparently, Marsh-marigolds can also be pollinated by rain. If the flowers are wide open when it rains, they collect large amounts of rainwater that drains off through the base of the perianth (because the petals are missing). Pollen is washed all over the wet flower's interior and pollination sometimes takes place.

Monk's-hood *Aconitum napellus*

Aconitum is a genus of more than 250 species, most of them native to the more mountainous regions of the Northern Hemisphere, often growing in mountain meadows. Monk's-hood, sometimes known as Wolfsbane, is a beautiful herbaceous perennial that is native to western and central Europe. In the UK it is a local native in shady sites by streams in the south-west and south Wales but it is widely naturalised elsewhere in similar habitat. It is also a common garden escape.

Monk's-hood and most other related species contain aconitine—a potent neurotoxin and cardiotoxin—and other alkaloids. They are extremely poisonous and have long been used for hunting using poison-tipped arrows, spears

Monk's-hood and Marmalade Hoverfly
Episyrphus balteatus

and harpoons to kill ibex, bears and even whales. Poisons derived from Monk's-hood relatives were also used in warfare by the Chinese. As poisonous plants with a history dating back to Greek mythology, Monk's-hood has featured in much historical and modern fiction, notably the books *Ulysses* by James Joyce and *I, Claudius* by Robert Graves; as well as the two television series, *Cadfael* and *Midsomer Murders*. Gardeners should remember that aconitine can be absorbed through broken skin. Be warned and take care.

Monk's-hood is a handsome plant that grows to a height of 150 cm or more with spikes of distinctive blue-violet flowers. Flowering takes place from May to June, sometimes continuing until September. Unlike most members of the buttercup family, Monk's-hood flowers have a complex bilaterally symmetrical structure with five sepals, the uppermost forming a large helmet-shaped hood which encloses the nectaries. The flowers can only be pollinated by long-tongued bumblebees with the strength to push their way into the flower to access the nectaries. But short-tongued bumblebees often chew their way into the flowers and steal the nectar. Monk's-hood is also visited by a few common pollen-eating hoverflies, notably the Marmalade Hoverfly *Episyrphus balteatus*.

Pasqueflower *Pulsatilla vulgaris*

The genus *Pulsatilla* includes about 33 species distributed across temperate areas of Europe, Asia and North America. Formerly, Pasqueflowers were included in the closely related genus *Anemone*. In the UK the Pasqueflower is a rare plant of dry calcareous grasslands, found mainly in the Chilterns, Cotswolds and East Anglia, including the Devil's Dyke where our photograph was taken. It is rare because the UK has lost more than 80% of its chalk grassland since the end of the Second World War. And it is becoming rarer because lack of grazing in crucial areas is leading to scrub encroachment.

Legend has it that Pasqueflowers grow on the blood-soaked graves of Roman or Viking warriors. In fact, they often do indeed grow on ancient burial mounds or barrows, probably because the barrows have often been left undisturbed for centuries.

All members of the genus *Pulsatilla* are very toxic. In American Medicinal Plants the listed symptoms of *Pulsatilla* poisoning include weeping and itching of the eyes, smarting and burning of the mouth and throat, sharp stomach and bowel pains, flatulence, frequent urge to urinate, tickling of the throat and coughing, and rheumatic pains, especially in the thighs.

The Pasqueflower is often regarded as one of the most gorgeous of the UK's native flowers. The flowers are solitary, 5–8 cm across. They lack petals but have six violet-purple sepals (covered with silky hairs on the underside), surrounding a dense cluster of golden stamens. Nectaries are located at the base of the outer stamens. Flowering begins in early spring, providing a useful early supply of pollen and nectar, and a few flowers are still available well into June. Pasqueflowers attract a rather generalist set of pollinators, including small bees, wasps, flies and beetles. They avoid self-pollination by having a stigma that is receptive only before the anthers shed their pollen. The flowers are also interesting in being heliotropic—they attract extra warmth by tracking the sun's movement so that they always face the sun.

Pasqueflowers produce attractive seed heads covered in long silky strands, making them resemble feather dusters. They persist for several months and the individual seeds, each with its silky plume, are eventually wind dispersed.

Pasqueflower

Columbine *Aquilegia vulgaris*

There are about 60–70 species of *Aquilegia*, all perennial plants found mainly at higher altitudes in much of the Northern Hemisphere. Columbine is found throughout much of Europe and is the only columbine native to the UK. It is a very common garden escape so its native distribution is now almost totally obscured. Columbine grows to about 80–100 cm tall and is locally present in woodland and meadows on calcium-rich soils, usually in edge areas with dappled shade. Garden escapes are more tolerant of soil type.

In the UK, Columbine is in flower from April to July. Native flowers are typically a beautiful deep blue but naturalised varieties and garden escapes are various shades of purple, mauve, pink or white. Columbines are elegant, pendant flowers with five curved spurs that secrete and store nectar. The most important legitimate pollinator is the Garden Bumblebee *Bombus hortorum*, a species with a tongue long enough to reach nectar deep in the spur. Columbine is often robbed by short-tongued bumblebees that gain access to the nectar by biting holes into the spurs. The worst culprits are usually Buff-tailed *Bombus terrestris* and White-tailed Bumblebees *Bombus lucorum*.

Columbines elsewhere in the world are visited by very different pollinators. In North America, for example, columbines have radiated and evolved different colours and different spur lengths to attract and cater for different pollinators, notably hawk-moths and hummingbirds. Blue columbines, like their Eurasian counterparts, are pollinated by bumblebees. Pale coloured, cream or yellow columbines differ in having extra-long spurs (15 cm long in *Aquilegia longissima* and are pollinated by nocturnal hawk-moths with long tongues. Several red columbines are pollinated primarily by hummingbirds. They secrete large volumes of nectar, with extra sugar content, and so satisfy the higher energy needs of hummingbirds compared with bees and moths.

Columbine and Garden Bumblebee *Bombus hortorum*

French Meadow-rue *Thalictrum aquilegiifolium*

Thalictrum is a genus of 120–200 species of herbaceous perennials native to temperate regions throughout most of the Northern Hemisphere. A few species are also found in southern Africa and South America. In the UK, there are three native species but several others, including this species, are popular as ornamentals and occur in the countryside as throw-outs or garden escapes.

French Meadow-rue, also called Siberian Meadow-rue or Greater Meadow-rue, is native to much of Europe and temperate Asia. In the UK, it is quite common in gardens and occurs locally in the wild as a throw-out in England and Scotland. It is a vigorous plant growing to about 1 m tall with dense clusters of fluffy pink or pale purple flowers. The individual flowers are small and lack petals. The sepals are also small and fall off soon after the flower opens. The long distance advertising directed at insect pollinators is provided by the numerous, long, colourful stamens. The paddle-shaped anthers are flattened and slightly wider than the filament. Though some meadow-rues are wind-pollinated, others including French Meadow-rue are pollinated by bees. French Meadow-rue flowers in late spring and early summer. The flowers do not produce nectar but have plentiful pollen that is collected by Honey Bees and bumblebees.

French Meadow-rue and Honey Bee *Apis mellifera*

Grossulariaceae (Currant family)

The family Grossulariaceae includes around 150–200 species, all in the genus *Ribes*. They are found throughout temperate regions of the Northern Hemisphere and many species range into the Southern Hemisphere in the mountains of Central and South America, particularly in montane and alpine habitats of the Andes above about 2,500 m.

Ribes includes numerous species that are cultivated for their fruits, notably Gooseberries, Black Currants, Red Currants and their hybrids. A few other species, including the Flowering Currant, are cultivated as garden ornamentals. In the past, North American Blackfoot Indians used blackcurrant root *Ribes hudsonianum* to treat kidney diseases and menstrual and menopausal problems. And Cree Indians used the fruit of *Ribes glandulosum* to assist women to become pregnant.

Species of *Ribes* are small deciduous shrubs. The flowers are mostly visited and pollinated by bees and a few other insects, though hummingbirds are the probable pollinators of some Neotropical species with large red flowers. Many birds eat the berries enthusiastically and undoubtedly disperse the seeds.

Flowering Currant *Ribes sanguineum*

The Flowering Currant is a native of the western United States and Canada but it has been widely cultivated and subsequently naturalised in much of Europe and Australasia. It was introduced into UK gardens in 1826, where it is now a popular garden shrub in the same genus *Ribes* as several other shrubs with soft fruits, including Black Currants *Ribes nigrum*, Red Currants *Ribes rubrum* and Gooseberries *Ribes uva-crispa*.

The Flowering Currant's berries are rather insipid and not highly regarded, though they were harvested and eaten raw, dried or stewed by the indigenous people of north-western North America. In more recent times, an exotic infusion of flowers and berries in vodka and honey is said to be very good.

The Flowering Currant is a deciduous shrub that grows to 2 m tall and broad. It makes a good hedge. It is particularly valuable because

Flowering Currant and Honey Bee *Apis mellifera*

it flowers in March, April and early May, providing an early source of nectar for wildlife. The plentiful flowers are borne in dangling clusters of up to 30 flowers. The flowers themselves are 5–10 mm in diameter with 5 pink or red petals. In their native North America, the flowers are visited and pollinated by hummingbirds attracted by a copious supply of nectar. In the UK, they attract mainly bees, including Honey Bees, bumblebees, flower bees *Anthophora* and a few of the larger solitary mining bees *Andrena* spp.

The fruit is a purple berry about 1 cm long and popular with numerous, small fruit-eating birds that defecate the seeds and so disperse them far and wide.

Crassulaceae (Stonecrop family)

The stonecrop family is found throughout much of the world but estimates of the number of genera and species differ depending on the authority, ranging from about 30 to 35 genera with 900 to 1,500 species. There are major centres of diversity in the south-western USA and Mexico, southern Africa, the Mediterranean basin and the Himalayas. In the UK, there are about eight native species in three genera (most are in *Sedum*) plus a few naturalised introductions and garden escapes. Stonecrops are tolerant of poor soils and draught, hardy and easy to grow, making them popular garden and house plants.

Most stonecrops are herbaceous perennials but there are some shrubs. Most species are typically found in semi-arid, sunny, rocky habitats. They have water-storing, succulent leaves and often have waxy leaf surfaces that reduce water loss. The flowers are bisexual and radially symmetrical, usually with 4–5 petals.

Butterfly Stonecrop *Hylotelephium spectabile*

Sedum is a large genus with anything from 400 to 600 species (though some species previously in *Sedum*, including Butterfly Stonecrop, are now allocated to *Hylotelephium*. Numerous species of *Sedum* are found in gardens and valued for their draught-tolerance and for providing good groundcover. Butterfly Stonecrop, a native of China and Korea, is a particularly common and popular garden plant. It is a herbaceous perennial and grows to about 40 cm tall. Its star-shaped pink to reddish-purple flowers are tightly clustered in a flattish or slightly domed head, up to 15 cm across. The flowers are protandrous—their stamens release pollen and wither before the stigma is receptive. There are numerous cultivars.

Butterfly Stonecrop flowers in autumn from about mid-September to well into November, sometimes later. As its name suggests, butterfly stonecrop is reputed to attract many butterflies. However, it is a late-flowering plant and in recent years, with unpredictable weather and declining butterfly numbers, butterflies have seldom been more than scarce, irregular visitors in our garden. Honey Bees visit more often, sometimes continuing well into October.

Butterfly Stonecrop and Honey Bee *Apis mellifera*

Fabaceae (Pea or Bean family)

The Fabaceae (formerly the Leguminaceae and plants are still commonly known as legumes) is a large and economically important family of flowering plants. The family is widely distributed and the third largest (behind the daisy and orchid families), with about 19,000 species in about 750 genera. Most species are herbaceous perennials but they also include annual herbs, many shrubs and trees (e.g. *Acacia*, *Mimosa* and *Cassia*), and a few rainforest giants that tower to heights of over 50 m in *Dipteryx odorata* in Central and South America and over 85 m in

Koompassia excelsa, an emergent tree in South East Asia. In the UK, there are about 60 native legumes. The majority are herbs but about half a dozen are small shrubs (e.g. several gorse species and broom). There are also numerous ancient introductions and many garden ornamentals, including trees, e.g. Laburnum *Laburnum anagyroides*.

Many legume species have root nodules that contain bacteria involved in nitrogen fixation. These bacteria (Rhizobia) have the ability to convert atmospheric nitrogen into a form of nitrogen that is usable by the host plant. This ability reduces fertiliser costs for legume crops and also means that legumes can be involved in crop rotations that replenish soils in which the nitrogen content has been depleted.

Legumes have been involved in human agriculture in both Eurasia and the Americas for at least 6,000 years. Many bean species became particularly important as staple foods, essential as rich sources of protein to balance the deficiency of the amino-acid lysine in cereal crops. Other important legume crops include lentils, peas, soya and peanuts; as well as forage crops, such as alfalfa and clover. Other farmed legumes include *Indigofera*—a source of the dye indigo—and numerous ornamental trees and shrubs, such as *Laburnum*, *Erythrina*, *Delonix*, *Acacia* and *Mimosa*. And a few legume trees are sources of valuable wood, notably *Dalbergia* (rosewood and cocobolo) and *Peltogyne* (purpleheart).

Worldwide, the majority of legumes are pollinated by bees, though there are tropical species that are pollinated by hawk-moths, birds, especially hummingbirds, and nectar-feeding bats. British legumes are pollinated almost exclusively by bees (see species accounts below).

Meadow Vetchling *Lathyrus pratensis*

Meadow Vetchling, also known as Meadow Pea or Meadow Pea-vine, is a native of Europe and Asia and has been introduced for use as animal fodder to other parts of the world, including North America.

Meadow Vetchling [PC]

Meadow Vetchling is easily the commonest and most widespread legume with tendrils that is found in lowland Britain. It also occurs sparsely up to about 450 m in mountainous regions. It is a plant of damp rough grassland, waste ground, hedgerows and roadside verges. It has a single pair of leaflets with tendrils but also has stipules that are large and leaf-like. With the aid of the tendrils, the squarish, angled stems scramble up and over other vegetation, reaching heights of up to 1.5 m in scrub and hedges. It also spreads easily, sometimes forming dense stands, by means of runner-like underground stems (rhizomes) that can reach lengths of 7 m or more.

Meadow Vetchling is a perennial with yellow flowers that is in bloom from May to August. The inflorescence has a long stem ending in a short spike of 5–12 yellow flowers. The individual flowers are about 10–18 mm long with a typical pea structure, including a 'standard', two 'wings' and a 'keel'. The flowers are visited and pollinated by bumblebees and solitary bees. The flowers are followed by seedpods that are conspicuously black and up to 40 mm long when ripe.

Common Bird's-foot-trefoil *Lotus corniculatus*

Common Bird's-foot-trefoil is a native in temperate Eurasia, including the UK, and North Africa. It has also been introduced to many other parts of the world, including North America,

Common Bird's-foot-trefoil, north Norfolk coast [PC]

Australia and elsewhere, where it is widely grown as a high quality forage plant in pastures or for hay or silage. In regions with poor soil, it is also used as an alternative to Alfalfa. In spite of its good qualities, Common Bird's-foot-trefoil is regarded as an invasive species in parts of North America and Australia, where it is planted along roads to help control erosion but later often spreads out of control into natural areas.

In the UK, Common Bird's-foot-trefoil is abundant (below about 1,000 m) almost throughout, growing in grassland, meadows, roadside verges or any other grassy areas. Locally, it also occurs on coastal shingle banks and sand dunes. The name 'bird's foot' refers to the appearance of the seed pods which look remarkably like their namesake. It has also earned the names Eggs and Bacon and Granny's Toenails, the latter evoking a rather disagreeable comparison with the plant's claw-like seed pods.

Common Bird's-foot-trefoil is a low-growing, sprawling, perennial plant that reaches heights of little more than 20–30 cm. It is a tough plant—even when growing in pastures and gardens, it can survive intensive grazing, trampling and mowing. It blooms from May or June until September and its clusters of red-tinged, yellow flowers produce plenty of nectar. Bumblebees and other bees are the commonest pollinators. Common Bird's-foot-trefoil is also an important food plant for the caterpillars of several butterflies and moths, including the Wood White *Leptidea sinapis*, Common Blue *Polyommatus icarus*, Silver-studded Blue *Plebejus argus*, Green Hairstreak *Callophrys rubi*, Dingy Skipper *Erynnis tages* and Six-spot Burnet *Zygaena filipendula*.

Ribbed Melilot *Melilotus officinalis*

Ribbed Melilot, sometimes called Yellow Melilot or Yellow Sweet-clover, is naturalised and widespread in lowland Britain south of the Tyne. Its native range is said to extend from central and southern Europe to central Asia. It has also been introduced to North America, Australia, New Zealand and elsewhere as a forage and nitrogen-fixing crop. In the UK, it is a plant of rough grassland, waste ground, and disturbed cultivation.

It is said that Ribbed Melilot arrived in the UK in about 1835, as seeds from America together with clover. However, it is also said that in Tudor times Ribbed Melilot was known as 'King's Clover' because it was a herb used by Henry VIII (i.e. 200 years earlier). The active ingredient is coumarin—a diuretic, anticoagulent and rodenticide (related to warfarin) and antiseptic that relieves fluid retention (oedema), poor blood circulation and numerous other ailments. Coumarin has a pleasant, sweet aroma which led to Ribbed Melilot being used in tobacco and speciality cheeses. However, coumarin is toxic in large doses so it should be used only in moderation.

Ribbed Melilot is a tall, biennial legume, growing to 2.5 m tall with erect, branching stems. Its long, slender spikes of yellow flowers are available from June until September. The flowers produce abundant nectar and pollen that attract Honey Bees, as well as many species of bumblebees and solitary bees. The flowers are said to produce excellent honey with a hint of a vanilla-like flavour. The seed pods that follow the flowers are brown with transverse ridges and only 3–5 mm long. Usually the pods contain only a single seed.

Ribbed Melilot [PC]

Red Clover *Trifolium pratense* and White Clover *Trifolium repens*

Both Red and White Clovers are native to much of Europe and Asia (and north-west Africa in the case of Red Clover) and both species have been widely cultivated and naturalised in many regions of the world, including North and South America, South Africa, New Zealand, Japan and elsewhere. In the UK, both species are common more or less throughout, below 500 m.

Both Red and White Clovers are widely grown as forage crops, valued because both are nitrogen fixers that improve soil fertility and reduce the need to use fertilisers to maintain productivity. They are also used as green manure crops.

Red Clover has been used in herbal medicine to treat respiratory problems (including asthma, bronchitis and whooping cough), eczema and psoriasis, and women's menopausal symptoms. However, it should be noted that research has not proved conclusively that the herb is effective for these or any other health concerns.

In the UK, there are more than a dozen species of *Trifolium* (clovers and trefoils) but the Red and White Clovers are particularly abundant and familiar. Both species are perennial plants found in all sorts of grassy places—pastures, meadows, lawns, parks and roadside verges. Both are quite variable in height but often reach 20–40 cm tall, with rounded flower heads composed of many tiny individual flowers (florets) packed so tightly together that they resemble a single flower. Both species flower from about June to September, each of them producing enough nectar to attract plenty of bumblebees and other bees of various species.

It must also be mentioned that there are clover farms in the USA that specialise in

White Clover and Clover Melitta *Melitta leporina* [PC]

producing four-leaf clovers that are sold as good luck charms. The idea that four-leaf clovers are lucky, and five-leaf clovers even luckier, has been around for well over a century. There are collectors who have searched for and found hundreds of four-leaf clovers. We have indulged in the pastime ourselves, particularly in Costa Rica and New Zealand, and have numerous fours and fivers to our name, several sixers and at least one eighter. But our finds are puny compared with those of a Japanese gentleman who once found a single stem of white clover with 56 leaves!

Gorse *Ulex europaeus*

There are about 20 species of gorse that are native to parts of western Europe and north-west Africa. Gorse stems are green, as are their branched spines, and together they are the main sites where photosynthesis takes place. There are three species of gorse in the UK—common Gorse which occurs almost throughout; Western Gorse *Ulex gallii* in the western half of the UK and along the East Anglian coast; and Dwarf Gorse *Ulex minor* in the south-east, south of the Thames.

Common Gorse is the most widespread, familiar species in the UK and the only one to be considered here. It is the largest species, reaching up to 3 m tall. The other two seldom reach much more than a metre in height. Common Gorse is typically found on dry, sandy or peaty soils, often forming extensive impenetrable thickets. Gorse is a fire-climax plant, both very flammable and well-adapted to survive fires. Burnt stumps sprout new growth from their roots and seed pods open and scatter seeds in response to fire, maintaining the seed bank and encouraging rapid regeneration. In years without fire, the seeds pods can be heard cracking open in hot sunshine and scattering their seeds. Like most legumes, Gorse has the capacity to fix atmospheric nitrogen, making it a useful plant for land reclamation where soil fertility has been depleted (e.g. mine tailings and abandoned, worked out mines).

Gorse can be found flowering throughout the year but looks its most showy in April, May and June. The golden-yellow, pea-type flowers have a distinctive coconut and vanilla scent and produce large amounts of pollen but very little nectar. The flowers attract many species of pollen-seeking bees, including bumblebees, Hairy-footed Flower Bees *Anthophora plumipes*, mining bees and others. Visiting bees force their way into newly opened flowers and trigger a spring-loaded pollination mechanism that releases pollen explosively and dusts the bee's abdomen.

Gorse also attracts other wildlife, particularly nesting birds. Its thorny thickets provide cover for the nests of many species. Birds that have a particular liking for Gorse include Dartford Warblers, Stonechats, Whinchats and Yellowhammers. The dense cover also provides essential shelter for Dartford Warblers in hard winters.

Gorse with Red-tailed Bumblebee *Bombus lapidarius* [PC] and Orange-tailed Mining Bee *Andrena haemorrhoa*

Rosaceae (Rose family)

Worldwide, the family Rosaceae consists of about 100 genera and 3,000 or more species, though the numbers vary according to different authorities. Members of the Rosaceae are found almost everywhere, except in Antarctica, but are most diverse in the Northern Hemisphere and rare in deserts and lowland tropical rainforest. Most members of the Rosaceae are woody shrubs or small trees, many of them armed with thorns or spines to deter herbivorous mammals. Others, including wild roses *Rosa* and blackberry relatives *Rubus*, are arching climbers or scramblers. Still others, including meadowsweet, cinquefoils and wild strawberries are herbaceous perennials. The family includes over 50 species in the UK.

The Rosaceae is a very important family economically—the source of many popular edible fruits, including apples, pears, medlars, quinces, loquats, peaches, apricots, plums, cherries, almonds, strawberries and raspberries. The family also includes numerous trees and shrubs, notably rowans, hawthorns, cotoneasters, firethorn and roses, that are planted in gardens as ornamentals. However, several of these ornamental trees and shrubs are now regarded as noxious weeds in some of the countries to which they have been introduced. Where relevant, details will be mentioned in species accounts.

Flowers of plants in the Rosaceae are radially symmetrical and bisexual. Most are white, pink or yellow and have five petals that form a shallow bowl and many stamens arranged in a spiral. The flowers vary from small to fairly large and their open-bowl shape provides easy access to nectar for a great variety of potential pollinators, including bees, hoverflies, butterflies and others.

Species of Rosaceae growing in our garden include several fruit trees, including apples, Crab Apples, pears, plums and cherries. All have flowers that attract a few pollinators but most are not particularly popular. The same applies to other family members in our garden, such as Rowan, Whitebeam and Hawthorn. Of course, all of these species have fruit that is very important as food for birds and other wildlife.

Blackthorn *Prunus spinosa*

Blackthorn is a native of Europe, western Asia and parts of north-west Africa and has been introduced to eastern North America, Tasmania and New Zealand. It is common throughout most of the UK in woodland edge, copses, scrub and hedgerows, sometimes becoming a tree up to six metres tall but more often seen as a dense spiny bush. Armed with savage thorns, Blackthorn is often used to make impenetrable, 'cattle-proof' hedges.

Blackthorn has several practical uses. The smallish blue-black, plum-like fruits (sloes) are harvested when ripe, usually in late October after the first frost of winter. They are

Blackthorn with Tawny Mining Bee *Andrena fulva* [PC] and Large Spotty-eyed Dronefly *Eristalinus aeneus*

incredibly astringent if eaten raw and are most commonly used to flavour sloe gin. Blackthorn wood polishes well and makes excellent walking sticks and tool handles. It is also the wood traditionally used by the Irish to make shillelaghs—the walking sticks or cudgels used to settle disputes, supposedly in a gentlemanly manner akin to duelling.

Blackthorn is a welcome, uplifting sight in early spring, in late March and April, when it bursts into a cloud of white blossom even before the new leaves unfurl. The white flowers, about 15 mm across, provide a valuable early source of nectar and pollen for bees and other insects. It is particularly attractive to a variety of mining bees *Andrena* and flies. Blackthorn is also important for other wildlife. The foliage is a food plant for the caterpillars of many moths and for the rare Black Hairstreak *Satyrium pruni* and Brown Hairstreak *Thecla betulae* butterflies; it also makes dense thorny thickets that provide safe cover for nesting birds; and in late autumn and winter its sloes are just small enough to be swallowed by diverse birds, mainly thrushes, including Blackbirds, Song and Mistle Thrushes, Fieldfares and Redwings.

Apple *Malus domestica*
The domesticated Apple originated in Central Asia, where its wild ancestor, *Malus sieversii*, is still found today. Apples have been grown for thousands of years in Asia and Europe, where winter Apples picked and stored in late autumn were an important winter food. Later, Apples were carried to North America by European colonists. The Apple is a very popular, edible fruit. Nowadays, there are said to be more than 7,500 varieties of Apples, and 2,500 in the UK alone, catering for various tastes and uses—for eating raw, for cooking or for cider production. In 2017, the worldwide Apple crop amounted to 83.1 million metric tons with about half grown in China.

In Victorian times, the UK grew more Apple varieties than anywhere else in the world— well over 2,000. Times have changed. Nowadays, the UK imports 70% of its Apples and only a few varieties make it onto supermarket shelves. However, there has been an encouraging surge of interest in having a better choice of UK Apples. Community orchards are springing up on unused land all across the country. With the help of volunteers, great efforts are being made to grow and ensure the survival of rare varieties, some of them with extraordinary names, notably Sheep's Head, Pig's Snout, Pig's Nose Pippin and Slack-ma-girdle.

In the UK the majority of Apple trees are found in commercial orchards and gardens. However, Apples, apple cores (often thrown from cars) and their seeds often escape and small naturalised trees can be seen growing on roadsides or in hedgerows, scrub and woodland. Apples are an important food source for wildlife in autumn and winter. They are eaten by many mammals and birds, particularly thrushes.

Apple blossom

A few Apple varieties are self-fertile but the majority are self-incompatible and have to cross-pollinate to produce fruit. Even those thought to be self-fertile do not do well on their own and fruit much better when pollinated by a bee or other insect. Researchers have studied fruit set and quality at harvest and found that, without insect pollination, the fruit harvest is reduced by nearly two thirds and is much poorer in quality, individual Apples being smaller and often misshapen. It has been estimated that insect pollinators contribute around £10,000 per hectare to Apple production.

Apple blossom puts on a spectacular display in late April and May. The five-petalled, white and pink flowers grow in clusters and attract a good variety of insects, especially bees and hoverflies. Our trees attract mainly bees, including a few bumblebees, plenty of Honey Bees, and several of the larger species of mining bees *Andrena*, including the Tawny Mining Bee *Andrena fulva* and Buffish Mining Bee *Andrena nigroaenea*.

Wall Cotoneaster *Cotoneaster horizontalis*

Wall Cotoneaster is a native of western China. It is one of more than a dozen species of cotoneaster that are now naturalised in many parts of the UK and often considered to be invasive (see below). Wild Cotoneaster *Cotoneaster cambricus* is our only native cotoneaster and is very rare.

Wall Cotoneaster and Buff-tailed Bumblebee
Bombus terrestris

Wall Cotoneaster is a popular ornamental shrub that grows vigorously almost anywhere, including shady, urban gardens, though it flowers more profusely in full sun. It tends to sprawl unless planted against a wall or fence. It grows to rather more than 1 m tall by about 1.5 m wide and is easily recognised by its tiny, glossy-green leaves and the characteristic, herringbone pattern of its stems and twigs.

Wall Cotoneaster flowers prolifically during April, May and June. The tiny, pinkish-white flowers are tubular but only 4–5 mm long. In spite of their small size, they are extraordinarily attractive to short-tongued bumblebees, such as Early *Bombus pratorum*, Tree *Bombus hypnorum*, Red-tailed *Bombus lapidarius* and Buff-tailed *Bombus terrestris* Bumblebees, particularly the tiny, first generation workers that are in the majority when Wall Cotoneaster is flowering. The flowers also attract a few Honey Bees and solitary bees.

Most species of cotoneaster, including Wall Cotoneaster, produce masses of bright red berries that ripen in autumn. Those of Wall Cotoneaster are a lot more popular than those of other cotoneaster species and provide an important source of food for birds, mainly thrushes and sometimes Waxwings, often well into the winter. How long they last depends on the extent to which they are defended by territorial birds, particularly Blackbirds. Because of their popularity with birds, the seeds are widely dispersed into the wild, where they are liable to compete with native vegetation. Several cotoneaster species are now considered to be invasive and are listed on Schedule 9 of the Wildlife and Countryside Act, meaning that it is *"an offence to plant or otherwise cause to grow these species in the wild"*.

Firethorn *Pyracantha coccinea*

Firethorn is one of seven species in the genus *Pyracantha*. It is a popular garden shrub that has been grown in Britain since the late 16th century and is now naturalised in southern areas. Its natural range extends from southern Europe to northern Iran, the Himalayas and China. It has been introduced as an ornamental to gardens in North America where it is often regarded as an invasive weed.

Firethorn is a thorny, evergreen shrub that can reach heights of 5 m and provides excellent cover and nesting sites for garden birds. It produces a prolific display of small white flowers in late May and June that attract huge numbers of small insect pollinators, mainly small bees *Andrena* and *Nomada* and small hoverflies *Syritta* and *Tropidia*. The flowers are too small to attract large bees or butterflies.

Small, long-lasting red berries are produced in autumn and early winter. Eaten raw by us, the berries are bitter and unpalatable but when cooked appropriately they make acceptable jams and marmalade. Because they persist for so long in winter, the berries are a valuable

Firethorn and Fork-jawed Nomad Bee
Nomada ruficornis

source of food for birds, particularly for Blackbirds and other thrushes. The birds disperse many seeds into wild areas. Note that the seeds are mildly poisonous in that they contain cyanogenic glycocides. But then so do the seeds of many other fruits found in the family Rosaceae, including apples, almonds, cherries and plums.

Hawthorn or May-tree *Crataegus monogyna*

Hawthorn is one of many rather similar species that are native to temperate regions of the Northern Hemisphere. In the UK there are two native species—Hawthorn and Midland Hawthorn *Crataegus laevigata*. They hybridise frequently so some plants can be difficult to distinguish. The common Hawthorn is abundant and widely distributed in lowland Britain below about 600 m.

If left to grow naturally, Hawthorns grow into small trees 10–15 m tall. More often, Hawthorns are used for hedges. In the 18th and 19th centuries, the Parliamentary Enclosures Act resulted in a surge of hedge planting—about 200,000 additional miles, most of it Hawthorns. It is estimated that there were still 500,000 miles of hedges in England at the end of the Second World War but only 236,000 miles by 1993 and less now. Hawthorn hedges are a mainly UK phenomenon and are extremely valuable for wildlife. The dense, thorny foliage provides protection and nesting sites for many birds; the leaves are a food plant for the caterpillars of numerous moths and other insects; the flowers are a rich source of pollen and nectar for small insects; and the red haws are important winter food for small mammals and thrushes, including wintering Fieldfares and Redwings.

Hawthorn with a Tapered Dronefly *Eristalis pertinax* [PC] and a dance fly *Empis tessellata*

In late April or May, Hawthorn trees and hedges burst into spectacular blossom. The white or pinkish-white flowers grow in umbel-like clusters. They are bisexual and have a sweetish scent mixed with a hint of rotting flesh. It has been shown that the latter odour is due to the presence of trimethylamine, a gas produced when carcasses begin to decompose. This association with death may be the reason it is considered bad luck to bring Hawthorn blossom into a house.

Hawthorn flowers are essentially self-incompatible and dependent on insects for pollination. Given their rather unpleasant scent, it is not surprising that flies of many species are among the most frequent visitors, along with solitary bees and a few butterflies and moths.

Bramble *Rubus fruticosus* agg.

Taxonomically, Bramble is a complex, or aggregate, of numerous microspecies. Its flowers are able to produce seeds from unfertilised ovules. Each microspecies differs slightly in such characters as the timing of flowering or fruiting, leaf size and shape, prickliness, flower colour, and fruit size or taste. Over 320 microspecies have been recognised in the UK and perhaps as many as 2,000 in Europe.

Bramble is a very common native species that occurs in most habitats throughout the UK, including woodland, scrub, heaths, hedges and grassland. Several factors contribute to its success—mammals and birds disperse its seeds efficiently, often over long distances; its seeds survive for several years in the seed bank; and it also spreads by vegetative growth, its long, arching stems frequently rooting wherever they touch the ground.

Bramble can become a nuisance in gardens and elsewhere. If conditions are favourable, it forms dense thickets that often outcompete other plants. However, if managed appropriately, it is a hugely desirable resource for wildlife. Its thickets provide cover for small mammals and nesting birds; bramble foliage is an important food plant for the larvae of many moths; its flowers are attractive to a wide range of pollinators; and the nutritious blackberries that follow are eaten by diverse birds and mammals; and also by us since at least Neolithic times (c. 8,000 BC).

Bramble and Honey Bee *Apis mellifera*

Bramble with Silver-washed Fritillary *Argynnis paphia* [PE] and Lesser Hornet Hoverfly *Volucella inanis*

Bramble is in flower from about June to September. The white or pinkish flowers provide plenty of easily accessible pollen and nectar. The five petals are often crinkly and form a shallow bowl that gives easy access to wide range of potential pollinators, notably Honey Bees, many different hoverflies, a few beetles, and numerous butterflies, especially Large White *Pieris brassicae*, Small White *Pieris rapae*, Meadow Brown *Maniola jurtina*, Gatekeeper *Pyronia tithonus*, Ringlet *Aphantopus hyperantus* and Silver-washed Fritillary *Argynnis paphia*.

Ripe blackberries are available from August through October. Blackberries are an aggregate fruit, composed of many tiny drupelets. They are popular with thrushes, Robins, *Sylvia* warblers (such as Blackcaps, Garden Warblers, Whitethroats, etc.) and Starlings. Blue Tits, Greenfinches and Bullfinches predate the seeds. Blackberries are also popular with mammals, including Badgers, Foxes, Grey Squirrels, Dormice and other rodents. Some flies visit overripe blackberries to imbibe the juice.

Dog-rose *Rosa canina* and other wild roses *Rosa* spp.

There are many species of wild rose, most of them rather similar, with white or pink flowers, curved thorns and red hips. There are a dozen or so native species and numerous hybrids that make identification difficult.

The Dog-rose is a good representative species that is native to Europe, north-west Africa and western Asia. In the UK it is the most widespread and abundant of our native roses, common more or less throughout, though more numerous in the south and sparsely distributed in northern Scotland. It is a deciduous scrambler that adorns hedgerows, woodland edge and scrub. Its curved thorns enable it to find purchase and to support itself amongst other vegetation. It regularly reaches heights of 3–5 metres but occasionally scrambles much higher.

The hips of the Dog-rose achieved celebrity and prominence in the UK during the second world war as a source of vitamin C. Made into rose-hip syrup, the hips are said to contain four times as much vitamin C as blackcurrant juice and 20 times as much as orange juice. As wartime children, we also remember that the hairy seeds inside rose hips make an excellent itching powder.

The sweet-scented flowers of the Dog-rose are very variable in colour, ranging from white through pale pink to deep pink. They have five petals and are 4–6 cm across. They are in flower in June and July, sometimes later, and attract mainly bumblebees, numerous solitary bees *Andrena* and *Lasioglossum* and flies.

Rose hips (about 15–20 mm long in the Dog-rose) redden round September and October but most remain hard and uneaten until late November and December. Rose hips are too large for many birds and Blackbirds and Fieldfares are the main consumers and seed dispersers. Many seeds are predated by Greenfinches, mice and voles.

Dog-rose and Common Carder Bumblebee *Bombus pascuorum*

Cucurbitaceae (Marrow and Cucumber family)

Worldwide, the Cucurbitaceae includes almost 100 genera and over 900 species. They are found in tropical and temperate regions but their range is limited by their sensitivity to freezing and near-freezing temperatures. Many members of the family are important foods and were among the first edible fruits to be cultivated in both the Old and New Worlds. Familiar examples include marrows, cucumbers, gourds, pumpkins, watermelons and melons. The family also includes luffas (or loofahs). If harvested when young, luffas are treated as a vegetable. Later, when fully ripe, luffas are very fibrous and used as a scrubbing sponge in baths and showers. Only one cucurbit species is native to the UK—White Bryony—though several other species, including melons, cucumbers and marrows, occur casually (e.g. in sewage farms and on rubbish tips), originating from human food waste.

White Bryony *Bryonia dioica*
White Bryony is native to much of Europe and northern Iran. It has also been introduced to parts of the USA, where it is classed as a noxious weed. It is the only cucumber relative that is native to the UK and occurs mainly in southern and eastern England.

White Bryony is very poisonous. The berries are poisonous to poultry but it is not clear whether the toxins are concentrated in the fruit pulp or the seeds, or both. Thrushes consume many berries without any ill-effects but probably defecate the seeds. The roots are also poisonous and occasionally get eaten by cattle with fatal consequences. White Bryony has bulky roots which, in the past, were sold as a counterfeit of mandrake roots *Mandragora*, a native of the Mediterranean region, and at that time popular as a painkiller, narcotic, hallucinogen and aphrodisiac. According to legend (recounted in the film *Harry Potter and the Chamber of Secrets*) when the root is dug up it screams and kills anyone who hears it.

White Bryony is a herbaceous, perennial vine. In the UK, it is common in hedgerows and woodland edge. Its stems reach up to 4–5 m long and it climbs using spirally coiled tendrils. It is dioecious with male and female flowers on separate plants and flowers for several months from about May to September. The greenish-white flowers are 12–18 mm across and very attractive to bees, including the Bryony Mining Bee *Andrena florea* and many other solitary bees (mainly species of *Lasioglossum*, and a few bumblebees and Honey Bees.

The White Bryony fruit is an ordinary-looking small, red berry, about 5–9 mm in diameter, nothing like the huge cucurbit fruits, such as marrows, cucumbers and melons. The first berries ripen in late July and continue ripening until September. Most are taken by *Sylvia* warblers, particularly Blackcaps, and some by thrushes.

White Bryony with a male Common Furrow Bee *Lasioglossum calceatum* and Common Yellow-face Bee *Hyalaeus communis* [PC]

Salicaceae (Willow and Poplar family)
In its traditional version, the willow family included the willows, poplars, cottonwoods and aspens. Now, following recommendations made by the Angiosperm Phylogeny Group based on evidence from DNA sequencing, the family has been greatly enlarged and now includes many species formerly in the Flacourtiaceae and the single species in the monotypic Scyphostegiaceae. The willow family now includes about 1,220 species in 56 genera, distributed more or less worldwide though the family is scarce in Australia and absent from New Zealand and Antarctica.

In the UK, there are more than 30 species or forms of willows *Salix* but this total includes numerous hybrid forms (sometimes involving three species) and a few ancient introductions.

Goat Willow *Salix caprea* and other willows *Salix* spp.

British willows range in size from large trees, over 30 m tall in the White Willow *Salix alba*, to small shrubs and even smaller shrubs. Our smallest willow, the Dwarf Willow *Salix herbacea*, is found on mountains in Scotland and Wales. It is a prostrate, creeping shrub, no more than 5–10 cm tall, and one of the smallest of all woody plants. All the British willows (with some called sallows or osiers) are typically found growing on damp ground often in wet woodland, on the banks of rivers or lakes, on boggy ground and, in the case of dwarf willows, on wet rocky ledges on mountain slopes.

Willows have many traditional craft and modern uses. Osier shoots or 'withies' (willow shoots) are so flexible that they can be woven into baskets, fish traps and wicker work, or to reinforce wattle-and-daub walls or fences. Willow was also used to make the oldest known fishing net. It was found in Finland, measured nearly 30 m by 1.5 m with a 6 cm mesh, and is dated to 8,540 BC. In more recent times, willows are often planted along stream banks to reduce soil erosion and also grown as an energy crop for use as biomass in power stations to generate electricity. Traditionally, willows were also used to relieve pain and the familiar product used nowadays—aspirin—is derived from salicin, a chemical compound found in the bark of willow trees.

Goat Willow, also known as Pussy Willow, is a good representative example of a willow and will be the main species discussed here. It is native to the UK and common, below about 750 m, in wet woodland and hedges almost throughout, and often found on drier ground than most other willows. It is a shrub or small tree reaching about 10 m tall and, compared with other species, has much broader, almost rounded leaves. Like other willows, the Goat Willow is dioecious—its male and female catkins are borne on different plants.

Goat Willow catkins are produced in early spring, usually before the leaves appear. Male catkins are silky and silver-grey when they first appear and become bright yellow as they mature in March and pollen becomes available. Female catkins are longer, thinner, green and offer nectar. The catkins, male and female, are an important source of early food for queen bumblebees, solitary bees and various flies. Although insects are very important as pollinators of willows, some pollen is also transported from catkin to catkin by wind.

The foliage of various willows is eaten by caterpillars of several moths and butterflies, including the Eyed Hawk-moth *Smerinthus ocellata*, Puss Moth *Cerura vinula* and, most notably, the Purple Emperor *Apatura iris*.

Goat Willow and Small Sallow Mining Bee *Andrena praecox* [PB]

Geraniaceae (Geranium family)

Worldwide, the Geraniaceae contains 7–11 genera and over 800 species distributed mainly in temperate and subtropical areas of the world, though a few are tropical. Most are herbaceous but a few are woody shrubs. The most important genera are *Geranium* (the crane's-bills, or true geraniums) with about 430 species; *Pelargonium* (the garden or house plants often called geraniums) with almost 300 species, and *Erodium* (stork's-bills) with about 80. *Pelargonium* has its centre of diversity in the Cape region of South Africa. In the UK, there are just the two genera in the wild—*Geranium* and *Erodium*.

The family has limited economic importance, though the genera *Geranium*, *Pelargonium* and *Erodium* have contributed numerous ornamental species and hybrids to horticulture. Also, several species of *Geranium* and *Pelargonium* are grown commercially for essential oils. Geranium oil, for example, is obtained from *Pelargonium odoratissimum* and relatives and is used in fragrances.

There are over 400 species of crane's-bills in *Geranium*, a genus widespread in temperate regions of the world, including tropical mountains, but absent from the humid tropics, deserts and polar regions. Note that the common name 'geranium' is also used for species of *Pelargonium*—popular house plants. The name 'crane's-bill' comes from the appearance of the seed capsule which resembles the bill of a crane or heron.

There are about a dozen native British crane's-bills, several more that are naturalised and many garden varieties and hybrids. Most crane's-bills are herbaceous perennials, in the range of 40–80 cm tall, and many spread vigorously and make excellent ground cover plants. If an appropriate selection of species or varieties is planted, it is possible to have crane's-bills in flower continuously from April through October, and so provide a reliable and popular source of nectar for a good variety of pollinators. Of the crane's-bills growing in our garden, for example, mostly non-natives and hybrid cultivars, the earliest is Rock Crane's-bill which flowers from mid-April through May. It is followed by two hybrid varieties—Purple Crane's-bill in June and July, and 'Rozanne' Crane's-bill from mid-June through August, much of September and sometimes October. Flowering latest is French crane's-bill from June until October or sometimes even November.

The majority of crane's-bill flowers come in shades of pink or blue. They have five petals forming a shallow bowl and range in size from less than 10 mm across in the smaller species up to about 30 mm in the largest. Many species have dark lines on their petals—nectar guides for pollinators—which converge on the nectaries, stamens and stigma at the centre of the flower. Crane's-bills with large flowers, such as Meadow Crane's-bill *Geranium pratense* and Bloody Crane's-bill, are almost always cross-pollinated. They show a variation of protandry in which their stamens, after shedding their pollen, bend away from the not yet receptive stigmas. Large-flowered crane's-bills are an important source of pollen and nectar and are pollinated mainly by bumblebees, Hairy-footed Flower Bees *Anthophora plumipes*, solitary bees, and many hoverflies. On the other hand, Cut-leaved Crane's-bill *Geranium dissectum* and Herb-Robert *Geranium robertianum*, both of which have small flowers, are automatically self-pollinated—their anthers and stigmas are in close proximity and mature at the same time.

Crane's-bills are plants that have explosive seed dispersal. As the seed capsules dry out, tension builds up in their tissues. When the capsules eventually split open, they do so explosively, ejecting seeds several metres at high speed.

French Crane's-bill *Geranium endressii*

French Crane's-bill is native to the western Pyrenees of France and Spain and a popular and widespread garden plant. In the UK, it is an often common and widely scattered escape from gardens and frequently found naturalised in the wild, on verges and waste ground, usually near houses.

French Crane's-bill is a sprawling plant that grows to about 60 cm tall, forming a dense mound of glossy foliage. It outcompetes many other garden flowers and so makes a very effective groundcover plant. It is also very hardy, surviving temperatures as low as -20°C. Its attractive flowers are 24–28 mm in diameter

French Crane's-bill and Hairy Yellow-face Bee *Hylaeus hyalinatus*

with 5 more or less unnotched petals, coloured salmon-pink with darker veins. French Crane's-bill has a very long flowering season, from late May or June and sometimes continuing through October into November. It attracts a good variety of pollinators, including several species of bumblebees, numerous solitary bees and a few butterflies.

'Rozanne' Crane's-bill *Geranium* 'Rozanne'

'Rozanne' Crane's-bill is a hybrid cross between *Geranium himalayense* and *Geranium wallichianum* (Buxton's variety). Both parent species are from the Himalayas. 'Rozanne' Crane's-bill began its existence in 1989 in the garden of Donald and Rozanne Waterer in Somerset. It then came onto the market in the 1990s and soon became a great success, winning numerous awards, including the Royal Horticultural Society's Award of Garden Merit in 2006 and the 'Plant of the Centenary' in 2013 in a competition celebrating 100 years of the RHS Chelsea Flower Show. Also, many garden designers include the 'Rozanne' Crane's-bill in their top ten perennial plants of all time.

'Rozanne' Crane's-bill and Large Narcissus Fly *Merodon equestris*

'Rozanne' Crane's-bill is a vigorous, hardy plant that forms dense, spreading mounds of foliage up to about 60 cm high, making it an excellent ground cover plant that reduces the need for weeding any competition. This celebrated crane's-bill produces masses of saucer-shaped violet-blue flowers, veined with purple, and with a white centre. The flowering season is very long, lasting for several months from late May or June until September or even October. It is a sterile hybrid and hence does not self-seed like many other crane's-bills. It does, however, produce nectar and pollen and attracts plenty of pollinators, especially hoverflies and a few bees and butterflies.

Bloody Crane's-bill *Geranium sanguineum*

Bloody Crane's-bill is a widespread native of Europe from southern Finland, extending east in temperate Asia as far as the Caucasus and Caspian Sea and south to the Mediterranean and parts of northern Africa. In the UK, it is locally distributed below about 1,200 m, particularly in northern England, usually growing in calcareous grassland, on chalk downland, in fissures of limestone pavement, and on cliffs and dunes near the sea. It is a showy and popular ornamental in gardens so, when found elsewhere close to houses, it is likely to be a garden escape or throw-out that often becomes naturalised. Our photograph was taken at the Devil's Dyke, adjacent to Newmarket racecourse—a chalk grassland Site of Special Scientific Interest that is home to several rare plants, including Pasqueflower *Pulsatilla vulgaris*, Purple Milk-vetch *Astragalus danicus* and Lizard Orchid *Himantoglossum hircinum* as well as Bloody Crane's-bill.

Bloody Crane's-bill is a clump-forming perennial that grows to a height of 30–40 cm. Its flowers are bright red-purple or magenta, veined with purple and 3–4 cm in diameter.

Bloody Crane's-bill and Swollen-thighed Beetle *Oedemera nobilis*

They are in flower for several months, from May until August or September, and attract a good diversity of insects, particularly bumblebees, hoverflies, butterflies and a few beetles.

Purple Crane's-bill *Geranium × magnificum*

Purple Crane's-bill is a sterile hybrid cross between *Geranium platypetalum* and *Geranium ibericum*. Both parents grow naturally in western Asia, including Turkey and the Caucasus. In the UK, Purple Crane's-bill is popular in gardens and any escapes or throwouts often persist and become naturalised on roadside verges and wasteland, mainly in the vicinity of villages and habitations.

Purple Crane's-bill is a hardy, clumpforming perennial that grows to about 70 cm high and in diameter. Its flowers are a darkveined, rich violet-blue and up to 5 cm across. The species blooms prolifically in early summer, in June–July, but has a much shorter flowering season than some other crane's-bills, notably the popular 'Rozanne' Crane's-bill. When flowering, it attracts many insects, especially bumblebees and a few hoverflies.

Purple Crane's-bill and Early Bumblebee *Bombus pratorum*

Rock, Big-root or Balkan Crane's-bill *Geranium macrorrhizum*

Rock Crane's-bill is a native of the Balkans in south-east Europe. However, it is a popular ornamental plant in British gardens, and elsewhere, and readily escapes into the wild where it persists close to habitations and often becomes naturalised, at least for a while. Rock Crane'sbill typically has purplish-pink flowers but there are many named garden cultivars with flowers ranging in colour from white though pale pink to magenta.

Rock Crane's-bill is a versatile perennial that thrives in full sun or even dry shade. It has underground rhizomes, enabling it to quickly form dense mats up to 40 cm high, making it an

Rock Crane's-bill and with a male Hairy-footed Flower Bee *Anthophora plumipes*

excellent groundcover plant that repels most competition. Its divided, toothed foliage is aromatic and sticky to the touch. The plants in our garden are in flower more or less throughout April and May. The pinkish-magenta flowers are about 3 cm across and have long, curved stamens, much longer than those of most other crane's-bills. They are mainly popular with long-tongued pollinators, including Common Carder Bumblebees *Bombus pascuorum*, Hairy-footed Flower Bees *Anthophora plumipes*, mason bees *Osmia* spp., and various flies, notably Common Snout-hoverflies *Rhingia campestris*.

Onagraceae (Willowherb and Evening-primrose family)

The family includes about 650 species in 18–20 genera. It is widespread, occurring on all continents except Antarctica and in regions ranging from tropical to boreal, though the greatest diversity is found in temperate areas of North, Central and South America. Most species including all the willowherbs *Epilobium* and evening-primroses *Oenothera*, are herbaceous though the fuchsias, which occur mainly in the Andes of South America, are woody shrubs or small trees.

The family includes a good number of popular garden ornamentals, notably the evening primroses and fuchsias. The evening-primroses are also of economic significance because evening-primrose oil is used extensively as a dietary supplement to treat various ills (see below).

The family has flowers with a structure built on a plan of four—four sepals, four petals and either four or eight stamens. Fuchsias are notable because their sepals enhance the flower's attractiveness by being as brightly coloured as the petals.

Most members of the family are pollinated by bees though moths, including hawk-moths, and flies are involved with some species. Hummingbirds are the most important pollinators of fuchsias in the Neotropics and honeyeaters in New Zealand. Most family members produce capsular fruits with seeds that are dispersed by ants, wind or water. The juicy, fleshy berries of fuchsias are dispersed by birds, especially tanagers.

Great Willowherb *Epilobium hirsutum*

The native range of Great Willowherb includes most of Europe as far north as southern Sweden, and parts of both Asia and North Africa. And it has been introduced to North America and Australia. In the UK, it is common in England, Wales and Northern Ireland, below about 400 m, but in Scotland it is confined to the far south and the east coast as far north as Caithness. It is typically found in damp habitats, including edges of marshes, rivers, ponds and ditches, and less often occurs on drier waste ground or roadside verges. It likes sunshine and is not shade tolerant.

Willowherbs can be difficult to identify because they hybridise freely but Great Willowherb is easier than most. It is the largest and most impressive of a dozen or so species of *Epilobium* found in the UK. It is a tall perennial, reaching heights of almost 2 m, with stems and leaves densely covered with soft hairs. Hence its alternative name Great Hairy Willowherb. Great Willowherb blooms from late June to September. Its flowers are bright

Great Willowherb [PC]

purplish-rose, 15–25 mm across, with four notched petals, eight stamens and a cross-shaped stigma with four lobes. The flowers attract mainly bumblebees, Honey Bees and hoverflies, together with a few butterflies.

Great Willowherb produces downy, plumed seeds that are carried long distances by the wind. As many as 80,000 seeds are produced by a single plant and they are usually among the first seeds to colonise bare or degraded ground. Also of interest, caterpillars of the Elephant Hawk-moth *Deilephila elpenor* feed on the foliage of Great Willowherb.

Rosebay Willowherb *Chamaenerion angustifolium*

Rosebay Willowherb is a common perennial, native throughout most of temperate Eurasia and North America, including large areas of boreal forest. It is an excellent example of a pioneer species—one that readily colonises disturbed ground. It is also known as Fireweed because it thrives on burnt ground after forest fires. In the UK, Rosebay Willowherb was long considered to be a scarce woodland species but is often now regarded as a ubiquitous weed. Its status first changed around the time that the UK's railway network expanded in the 19th century and the change accelerated during the first and second world wars, when it proliferated wherever woodland had been cleared for the war effort and on city bomb sites. In London, during the Blitz, it was known as 'Bombweed'. Nowadays, in late summer, Rosebay Willowherb provides lovely drifts of colour on railway embankments, waste ground and along woodland rides.

In other parts of its range, but not the UK, Rosebay Willowherb is used in numerous ways. It can be used to treat infected cuts or boils. In Alaska it is used to flavour candies, syrups and ice cream. And in Russia, before tea from China became available, its leaves were used to make a tea-like drink. It was even exported to Western Europe as 'Russian Tea' or 'Ivan Chai'.

Rosebay Willowherb grows to about 1.5 m tall and has a long flowering season, from June to September. Its eye-catching spikes of rose-purple flowers attract bumblebees and Honey Bees—the main pollinators—as well as hoverflies and a few butterflies. The bisexual flowers have four petals, eight stamens and a four-lobed stigma. They are self-compatible, though their anthers shed their pollen before the stigmas are receptive, so self-pollination is largely avoided.

The fruits are capsules, 5–8 cm long, that split into four sections, releasing wind-dispersed, plumed seeds. Each capsule releases 300–500 seeds and the total seeds per plant can reach as many as 80,000.

Rosebay Willowherb and with Marmalade Hoverfly *Episyrphus balteatus*

Evening-primroses *Oenothera* spp.

The evening-primroses in the genus *Oenothera* include about 145 herbaceous species native to the Americas. They probably originated in Mexico and Central America and subsequently diversified and spread north and south. Several species are popular garden flowers and now naturalised in most temperate regions of the world. Evening-primroses were first introduced to Britain in the early 17th century and several are now widely naturalised. Evening-primroses interbreed quite easily. In the wild there are many hybrids and they sometimes form extensive hybrid swarms.

In the UK, two species—Common *Oenothera biennis* and Large-flowered *Oenothera glazioviana* Evening-primroses—are the most widespread. The Fragrant Evening-primrose *Oenothera stricta* is smaller and mostly found in sandy, coastal areas. Several other species occur casually but are often difficult to identify. In the wild, evening-primroses are quick to colonise cleared or disturbed areas, including overgrazed grassland, roadsides, railway lines, wasteland and dunes.

Evening-primroses are sometimes eaten. Young leaves can be cooked like spinach and the boiled roots make an acceptable potato-like vegetable. However, evening-primroses have achieved greater popularity as a 'cure-all' for an extraordinary list of disorders. As a herbal remedy, evening-primroses are used to treat eczema, sore throats and coughs, stomach aches, arthritis and, prepared as a poultice, bruises. More recently, evening primrose oil, extracted from the Common Evening-primrose, has been used as a dietary supplement to treat dermatitis, asthma, coughs, migraine headaches, diabetes, heart disease and numerous other afflictions. However, the effectiveness of evening-primrose oil is controversial and has been described as *"a remedy for which there is no disease"*. Claims for its beneficial effects are now largely discredited.

The Large-flowered Evening-primrose is the commonest evening-primrose in the wild in the UK and very similar to the others. We will treat it here as a typical representative of an evening-primrose. It is a biennial that reaches about 1.8 m tall and flowers from June to September. The flowers are large, 50–80 mm wide, and fragrant with four pale yellow petals. As their name suggests, the flowers open in the evening and remain open until about mid-morning the following day. Evening-primroses are visited and pollinated by various moths, including hawk-moths, and also by bees and other insects. However, their flowers (in common with those of members of a few other families) produce multiple pollen grains that are held together in loose groups by sticky (viscin) threads. Only insects that are appropriately specialised can collect such sticky strings of pollen grains and transfer them to other flowers effectively. It has been suggested that plants connect pollen grains together in this way as a response to a scarcity of pollinators. Having multiple pollen grains on a sticky thread ensures that any visiting pollinator will depart with more pollen than it would otherwise have done.

Evening-primrose plants produce enormous numbers of seeds—often over 100 capsules, each probably containing over 150 seeds. The seeds are a valuable food resource for seed-eating birds.

Large-flowered Evening-primrose *Oenothera glazkoviana* [PC]
and with Marmalade Hoverfly *Episyrphus balteatus*

Malvaceae (Mallow family)

As a result of DNA studies by the Angiosperm Phylogeny Group, the Malvaceae has been much enlarged and now includes various genera that were formerly included in the families Bombacaceae, Tiliaceae, and Sterculiaceae. The family is now estimated to contain at least 4,225 species, including herbs, shrubs and trees, in about 244 genera. The family has an almost worldwide distribution, except for very cold regions, and is most species diverse and numerous in the tropics.

By far the most important species economically are four species of cotton (grown in both the Old and New Worlds for more than 5,000 years), followed by cacao (the source of cocoa and chocolate), kola nuts (chewed for their caffeine content and also used to flavour beverages) and okra (a vegetable also called 'ladies' fingers'). There are also several genera which include popular garden ornamentals, the most familiar being *Hibiscus*, *Alcea* (hollyhocks), *Lavatera* (tree-mallows) and *Malva* (mallows).

The enlarged family also includes spectacular trees—notably the African baobabs and the South East Asian durians—all formerly included in the family Bombacaceae. Durian fruits weigh up to 8 kilograms and have a tough, spine-covered rind which protects numerous seeds surrounded by pulpy arils. Durians are regarded as a great delicacy but have a smell so overpowering that they are often banned from hotels, public buildings, and public transport. The great English naturalist Alfred Russel Wallace described the pulp as *"a rich custard highly flavoured with almonds."* Wallace was a fan (as are we) and described the sensation of eating durians as being *"worth a voyage to the East to experience."* Others have likened the experience to *"eating Limburger cheese in an outhouse."*

There are only six native members of the Malvaceae in the UK—three species of lime trees *Tilia* (formerly in the Tiliaceae), Musk-mallow, Tree-mallow *Malva arborea* and Marsh-mallow *Althaea officinalis*. Three others—Common Mallow, Dwarf Mallow *Malva neglecta* and Hollyhock *Alcea rosea*—are ancient introductions. And numerous other non-natives can be found as garden escapes or throw-outs.

The stems and leaves of all mallows contain mucilage which makes them slimy when crushed. Medicinally, the mucilaginous quality of mallows may be used like *Aloe vera* for soothing sunburns and other inflamed skin conditions, or internally for soothing coughs and sore throats. Of the native British species, the Marsh-mallow has an additional distinction as the plant originally used to make the sweet with same name. Ancient Egyptians are said to have been the first to make marshmallows, as long ago as 2,000 BC. The custom later spread to Europe. Ground up pieces of root, containing starch, sugars, oils and mucilage were first boiled. The remaining liquid and added sugar was then beaten to a froth that thickened when cool. Modern marshmallows lack the root and are entirely artificial.

The flowers of mallows and hollyhocks have funnel-shaped flowers with five petals and a tightly packed column of stamens surrounding the pistil. The flowers are bisexual but avoid self-fertilisation by having anthers that shed pollen before the stigmas are receptive. Honey Bees, bumblebees, hoverflies and pierid butterflies are the most frequent visitors.

Musk-mallow *Malva moschata*

Musk-mallow is native to Europe from Spain north to the UK and Poland, and east to Turkey and southern Russia. In the UK, it is common in most areas but rare or absent in much of Scotland and Northern Ireland. Musk-mallow is a plant of pastures, field margins, hedgerows and roadside verges. It is also a popular and attractive plant much planted in cottage gardens and in newly sown perennial meadows. It mixes well with other perennial meadow plants, such as Meadow Crane's-bill *Geranium pratense*, Field Scabious *Knautia arvensis*, knapweeds *Centaurea* spp., Viper's-bugloss *Echium vulgare* and Wild Marjoram *Origanum vulgare*.

Musk-mallow is a herbaceous perennial, with hairy stems, deeply cut upper leaves and heart-shaped lower leaves. It grows to about 60–80 cm tall and its showy flowers, about 60 mm across, grow in loose clusters and are coloured a delicate, pale rose-pink. As suggested by its name, the

Musk-mallow [PC] and with a solitary bee

flowers have an alluring musky scent that is most noticeable in warm weather or indoors. Musk-mallow has a long flowering season, first appearing in late June or July and continuing until late August or September. The flowers avoid self-pollination by having anthers that mature and shed their pollen before the stigma has become receptive. The flowers are less attractive to pollinators than those of Common Mallow and Hollyhocks but do attract a good variety of insects in small numbers, including bumblebees, solitary bees, hoverflies and butterflies.

Common Mallow *Malva sylvestris*

Common Mallow is native to much of Eurasia and North Africa but is now naturalised in many parts of the world, including the USA, Canada, Mexico and Australia. It was introduced to Britain long ago, perhaps by the Romans, and is a widespread and common plant of roadsides, hedgerows and waste ground throughout much of the lowlands.

Young leaves, shoots and flowers of Common Mallow were eaten as vegetables by Romans garrisoned in Britain but have long been even more popular as a herbal remedy for almost everything—kidney, bronchial, digestive and skin problems, fevers, insect stings, wounds,

Common Mallow and with a Honey Bee *Apis mellifera*

inflammation and even stupidity. The stems and leaves of Common Mallow have a high mucilage content with soothing emollient properties.

In the UK Common Mallow is a vigorous perennial that thrives in open, sunny areas, growing to a height of about a metre. Sometimes it is erect and almost bush-like but often it scrambles and sprawls over other vegetation. Common Mallow flowers from June to September. Its attractive flowers have five notched purplish-pink petals with conspicuous darker stripes. The bisexual flowers avoid self-fertilisation by having anthers that shed their pollen before the stigmas are receptive. Honey Bees and bumblebees are abundant visitors along with smaller numbers of solitary bees, hoverflies and occasional butterflies. Bees, especially Honey Bees, often become smothered with pollen but apparently make little attempt to collect it in their pollen baskets. The pollen grains are large, spherical and covered with tiny, sharp spikes. Presumably, the spiky grains are too difficult for the bees to deal with.

The seeds or nutlets are wedge-shaped and tightly arranged in a ring like the segments of an orange or tangerine. The wedges are distinctly cheese-shaped and hence the local childrens' name 'Fairy Cheeses'. Because Common Mallow quickly succumbs to damage when it finishes flowering, it also goes by the picturesque name Rags and Tatters.

Hollyhock *Alcea rosea*

There are about 60 species of hollyhocks in *Alcea*, a genus that is native to Asia and Europe. Our Hollyhock is a classic cottage garden flower that was imported into Europe from China, probably in the 15th century but perhaps even earlier. There are many cultivars and it is a frequent garden escape, often appearing on common ground or railway embankments.

The Hollyhock is a biennial with a tall flowering spike that reaches heights of two metres or more. The flowers have five petals that overlap to form an open, funnel-shaped flower 70–100 mm across. They are very variable in colour, ranging from white through beige, yellow and pink to very dark red. The Hollyhock is what is known as a 'revolver flower'—nectar is accessed through a ring of narrow tubes visible around the base of the flower. To obtain all the available nectar, an insect has to circle and probe all the access tubes. Bumblebees are the main visitors and collect almost all the nectar. Surprisingly, they tend to ignore the abundant pollen but often become smothered by it as they move around inside the flower. The pollen grains are large and spiny which helps them to adhere to furry bumblebees but probably makes them difficult to manage.

Hollyhock and Common Carder Bumblebee *Bombus pascuorum*

Cistaceae (Rock-rose family)

Worldwide, the Cistaceae is a small family that includes eight or nine genera and about 180 species. All but a few herbaceous species are small aromatic shrubs that are widely distributed in warm-temperate regions. The main centre of diversity is the Mediterranean basin but the family also occurs as far east as Central Asia and in North, Central and South America. Important genera include the widely distributed *Helianthemum* with over 100 species; the mainly Mediterranean *Cistus* with 18 species; and *Lechea* with 17 species confined to the New World. Rock-roses are draught-tolerant and prefer dry, sunny habitats, though some are hardy and do well as ornamentals even in the cold winters of northern Europe.

In the UK, three species of *Helianthemum* and one of *Tuberaria* are native but only the Common Rock-rose is at all widespread and common. Numerous species of *Cistus* occur as garden ornamentals.

Many rock-roses have interesting and unusual adaptations. Most have the ability to create symbiotic relationships with fungi. The fungus colonises the root system of the host plant, augmenting its ability to absorb water and nutrients, while the host plant provides the fungus with carbohydrates formed by photosynthesis. The relationship allows the plant to thrive growing in poor soils where it might otherwise struggle.

Shrubs in the genus *Cistus* are also interesting because they have adaptations that enable them to survive wildfires. Their seeds have a hard, water-resistant coating and can remain dormant for many years, creating a large seed bank. As well as destroying competing vegetation, periodic fires crack the hard coating of the seeds, enabling them to germinate when next it rains. Large numbers of young shoots appear simultaneously and are likely to outcompete any other plants attempting to repopulate the burnt area.

Rock-roses are spectacular when flowering and are often covered profusely with showy flowers that are pink, yellow or white. The flowers are bisexual with five petals and numerous stamens. They are ephemeral and open only in full sunlight and remain open for only a few hours before they wither. The flowers produce little or no nectar but are attractive to a variety of pollen-eating insect pollinators, mainly bees and some hoverflies.

Common Rock-rose *Helianthemum nummularium*

The Common Rock-rose is native to much of Europe and the only member of the family that is frequent and widespread in the UK throughout much of England, Wales and Scotland. In southern England it is restricted to dry calcareous grassland but in Scotland it is also found on slightly acid soils, even growing alongside Heather on heaths. It thrives in sunny south-facing grassland, including rocky slopes and even near-vertical cliffs.

The Common Rock-rose flowers abundantly from June to September. The flowers are 15–25 mm in diameter and each has five sulphur-yellow, crinkly petals, sometimes with an orange spot on each petal, and numerous stamens. The latter are sensitive to being touched and splay apart when visited by a pollinator, revealing the stigma. Garden cultivars are found in a variety of colours, ranging from white though yellow to red. The Common Rock-rose provides enough pollen (but no nectar) to attract numerous wild bees and Honey Bees. It is also a food plant for the

Common Rock-rose [PC]

caterpillars of several butterflies, including the Green Hairstreak *Callophrys rubi*, Brown Argus *Aricia agestis*, Northern Brown Argus *Aricia artaxerxes* and Silver-studded Blue *Plebejus argus*.

Pink or Cretan Rock-rose *Cistus creticus*

The Pink or Cretan Rock-rose is a native of the eastern Mediterranean (but not endemic to Crete) and widespread in scrub and open woodland below about 1,200 metres. It is a very decorative evergreen shrub with grey-green foliage and abundant pink flowers, reaching about 1 m high. It thrives even in poor, well-drained soils but needs plenty of sun and can tolerate draught.

The bisexual flowers are 4–5 cm across with five crinkly, deep pink petals surrounding a cluster of golden stamens. The flowering season is long, from about mid-May until mid-August, and a few flowers continue to appear until the end of September or even November in mild years. The flowers open only in warm sunshine and last for only a few hours before withering. They are pollen-only flowers that lack nectar. In our garden, they attract bumblebees, particularly Tree Bumblebees *Bombus hypnorum* and Common Carder Bumblebees *Bombus pascuorum*, but also attract even more of the short-tongued hoverflies that eat pollen as well nectar. They also attract a few greenbottles *Lucilia* spp. Not having nectar, they do not attract any butterflies.

Pink Rock-rose with Tiger Hoverfly *Helophilus pendulus* and Large Narcissus Fly *Merodon equestris*

Gum Rock-rose *Cistus ladanifer*

The Gum Rock-rose is native to the western Mediterranean, including the Iberian peninsula, southern France and north-west Africa. It is well adapted to the Mediterranean climate with its long-summer droughts and cold winters. It is an aggressive plant that has invaded farmland and grassland in the mountainous regions of central Spain and southern Portugal.

Gum Rock-rose and Batman Hoverfly *Myathropa florea*

The Gum Rock-rose is an evergreen shrub that sometimes reaches more than 2 m tall and wide. Like many rock-roses, it forms symbiotic mycorrhizal associations with various fungi. The fungi improve the rock-rose's ability to absorb water and nutrients while, in return, it absorbs carbohydrates from the host plant. The latter is enabled to thrive even in poor, dry soils. The foliage of the Gum Rock-rose is covered with a sticky brown resin, known as labdanum, that has an aromatic balsam smell and is used in the perfume industry. It is valued as a substitute for ambergris, a product which is now largely banned from use because it is obtained from Sperm Whales—a protected, endangered species.

In the UK, the Gum Rock-rose is a popular ornamental, appreciated for its scented foliage and spectacular white flowers, 5–8 cm in diameter, with a dark red blotch at the base of each petal. It flowers for 5–6 weeks in June and July. The flowers reward pollinators with pollen, not nectar, and so attract bees and some hoverflies, but not butterflies.

Limnanthaceae (Meadow-foam family)

The meadow-foam family is very small, consisting of only two genera and eight species that occur in temperate North America. All species are annual herbaceous plants. The monotypic genus *Floerkea* inhabits shaded, ephemeral pools in deciduous forests in eastern North America and humid coniferous forests along the north-west coast. The seven species of *Limnanthes* are native to western North America and are typically found around the margins of seasonal wetlands. White Meadow-foam *Limnanthes alba* is grown commercially for the oil and honey made by bees from its flowers. It is famous for its distinct flavour of toasted marshmallows. And the Poached Egg Plant, which has pretty white and yellow flowers, is widely cultivated in gardens in both North America and Europe.

Meadow-foam *Limnanthes douglasii*

Meadow-foam is a native of California and Oregon in the western USA, where it grows alongside seasonal ponds or in other wet grassy habitats. It was first collected by David Douglas, a Scottish explorer and botanist who worked in the west coast of America in the 1820s. It is a popular ornamental in gardens and a recipient of the Royal Horticultural Society's Award of Garden Merit. As an ornamental, it is often known as the Poached Egg Plant, because the flower, with its yellow centre surrounded by white, resembles a poached egg. It is a frequent garden escape on roadsides and tips, and often persists in suitable areas.

Meadow-foam is a hardy, easy-to-grow annual that prefers full sun and reaches a height of up to 20–30 cm. The bright and cheerful, fragrant flowers bloom from June to September and attract bees and many hoverflies. Its popularity with hoverflies makes meadow-foam an excellent companion plant for vegetable crops because hoverfly larvae are important predators of aphids, thrips, small caterpillars and other soft bodied pests. Meadow-foam self seeds easily.

Meadow-foam

Brassicaceae (Mustard and Cabbage family)

The Brassicaceae is a fairly large and economically important family of flowering plants, most of them herbaceous. Estimates of the size of the family vary but are in the region of 3,700–4,000 species arranged in 340–370 genera. The family has a worldwide distribution, excluding Antarctica, and is particularly diverse in temperate areas. The family is also known as the Cruciferae based on the fact that the flowers have four petals arranged in the shape of a cross.

The genus *Brassica*, which is native to western Europe, the Mediterranean region and temperate areas of Asia, includes many plants grown as vegetables that are especially nutritious and of great economic importance. There is some evidence that diets rich in cruciferous vegetables lower the risk of developing various types of cancer. *Brassica* crops have been grown in Europe for 8,000 years and some have been fundamentally altered and domesticated by selective breeding. For example, different varieties of the species *Brassica oleracea* now include such familiar, but different, vegetables as Cabbage, Broccoli, Brussels Sprouts, Cauliflower, Kale, Kohlrabi and Collard Greens. Other important *Brassica* species include *Brassica rapa* (Turnip and Chinese Cabbage) and *Brassica napus* (Oil-seed Rape). The latter is now a very common roadside escape from cultivation. The family also includes Watercress *Nasturtium officinale* and several species used as culinary condiments, including Horse-radish *Armoracia rusticana*, White Mustard *Sinapis alba* and Black Mustard *Brassica nigra*. White Mustard is an ancient introduction to the UK but Black Mustard may be native. Both species are fairly common in southern Britain and the midlands.

Woad *Isatis tinctoria* is yet another famous brassicaceous plant. It is a biennial or perennial herb with pretty yellow flowers, native to much of Eurasia and naturalised in western North America where it is regarded as a noxious weed. Woad was an important source of a natural blue dye—indigo—and was cultivated in much of Europe. In medieval Britain it is said to have been used by Ancient Britons to paint patterns on their bodies. And, according to Richard Mabey in *Flora Britannica*, the ammoniacal stench produced by the manufacturing process was so disgusting that Elizabeth I banned the processing of Woad in any town though which she was passing. Eventually, in the 12th and 13th centuries, Woad was replaced by true Indigo *Indigofera tinctoria*, a more colourfast import from India. In the early 20th century both were replaced by synthetic blue dyes.

The Brassicaceae is also the source of several popular ornamental plants, notably Aubretia, Evergreen Candytuft, Dame's-violet *Hesperis matronalis*, Honesty, Sweet Alison *Lobularia maritima* and Wallflower, most of which attract a useful selection of pollinators.

Wallflower *Erysimum cheiri*

Worldwide, there are more than 150 species of wallflowers, native to temperate areas of Europe, the Mediterranean region, south-west Asia, and both North and Central America. Many species are endemic to very small areas, such as small islands, mountain ranges or isolated volcanoes.

Many species of wallflower have a long history of use in traditional medicine, though medical uses became uncommon in Europe after the Middle Ages. Despite concerns about its safety, people occasionally use Wallflower for heart problems. Like the Foxglove *Digitalis purpurea*, it is useful in small doses but very toxic in larger doses. The active ingredients include cardiac glycosides and other toxic phytochemicals that are important to the plants for defence, protecting them against insect herbivory.

In the UK, most Wallflower garden varieties are derived from *Erysimum cheiri*, a native of southern Europe. The species was first cultivated in Britain in medieval times and recorded in the wild as early as 1548. Nowadays, it is fairly common throughout much of lowland Britain, mainly in southern areas. It is most often found growing in dry, sunny habitats, including railway cuttings, quarries, sea cliffs, ruined buildings and on old walls.

Wallflower is a popular evergreen, perennial garden plant that grows to a height of 50–60 cm. It has a long flowering season, from early March until June or sometimes July. The fragrant flowers are variable in colour but mainly bright yellow, orange or dull red. The sepals form a tube-like entry to the nectaries, providing access to long-tongued insects, including bees, hoverflies and butterflies. The most frequent visitors in our garden are the Hairy-footed Flower Bee *Anthophora plumipes*, Common Carder Bumblebee *Bombus pascuorum*, Dark-edged Bee-fly *Bombylius major*, Common Snout-hoverfly *Rhingia campestris* and various pierid (whites) butterflies.

Wallflower and a male Hairy-footed Flower Bee *Anthophora plumipes*

Honesty *Lunaria annua*

Honesty is a native of the Balkans and south-west Asia but has long been a popular garden plant and is now naturalised more or less throughout temperate regions of the world. In the UK it has been grown in gardens since the mid-16th century and since then has become a common garden escape, usually near habitations, on roadsides, rubbish tips and waste ground throughout much of England and Wales and parts of Scotland.

Honesty has attractive flowers but is best known for its distinctive seedpods which measure about 30–60 mm across. The Latin generic name means 'moon-shaped'—a fitting description for the translucent pods which have something of the appearance of silvery coins. Hence the names Silver Dollars in the USA and Money Plant in South East Asia. The translucence of the

Honesty and a male Orange-tip *Anthocharis cardamines*

pods is also thought to be the origin of the English name Honesty. The seedpods persist on plants throughout the winter and are much prized for use in dried-flower arrangements.

Honesty is usually a biennial, growing to around 60–100 cm tall and does well in both dappled shady sites and sunshine. It is in flower from late March or April to late May or June. The flowers are reddish-purple, or sometimes white, and attract a good range of bees, flies and butterflies. Early on, in late March and April, the flowers are particularly attractive to Dark-edged Bee-flies *Bombylius major* and early emerging butterflies, such as Orange-tips *Anthocharis cardamines* and Brimstones *Gonepteryx rhamni*. Later, they also attract a few bumblebees and many hoverflies.

Aubretia *Aubrieta deltoidea*

Aubretia is one of about 20 species in the genus *Aubrieta* that occurs from southern Europe east to central Asia. Aubretia is very popular in gardens and a common garden escape. The genus is named in honour of Claude Aubriet, a French botanical illustrator who worked at the 'Jardin du Roi'— the royal botanical garden in Paris.

Aubretia is an evergreen, hardy, draught-tolerant, low-spreading perennial that forms dense mats and does best when growing in full sun. It is an excellent plant for groundcover and does particularly well in dry conditions. This makes it an ideal plant for rockeries or trailing

Aubretia and Dark-edged Bee-fly
Bombylius major

over old walls, where it brings welcome colour to gardens in early spring. Cultivated varieties range in colour from purple through lilac to pale pink and the cruciform flowers are about 20 mm across. Aubretia is an early flowerer, blooming from March to May, making it a valuable source of nectar for any insects on the wing early in the year. It is not strongly scented but is, nevertheless, a particular favourite of Dark-edged Bee-flies *Bombylius major*. In our garden, it also gets a few visits from miscellaneous bumblebees and solitary bees but usually flowers too early to attract many butterflies or hoverflies.

Evergreen or Perennial Candytuft *Iberis sempervirens*

Evergreen Candytuft is native to southern Europe, including France, Spain, Italy and the Balkans. It is also native to Morocco and Algeria in north-west Africa and Syria in the Middle East, and it is naturalised in both North and South America. In the UK, Evergreen Candytuft is a popular ornamental and occasional garden escape.

Evergreen Candytuft is a low, spreading shrub, growing to about 30 cm in height, and often grown in rock gardens. It is evergreen, as indicated by its specific name 'sempervirens' which means 'always green'. It likes sunshine and heat but is also frost hardy. It looks its best when cascading over rocks or walls and also makes a good groundcover plant.

Evergreen Candytuft continues to flower for several weeks during spring and early summer. The fragrant flowers are pure white in dense clusters, 4–5 cm wide. The individual flowers have distinctive large outer petals and are very attractive to various pollinators, especially solitary bees and hoverflies.

Evergreen Candytuft and
Tiger Hoverfly *Helophilus pendulus*

Caryophyllaceae (Campion and Pink family)

The Caryophyllaceae is a large family with over 80 genera and well over 2,000 species of mostly herbaceous plants. The majority are found in northern temperate areas, with just a few on tropical mountains and in the Southern Hemisphere. The most important centre of diversity is the Mediterranean basin and adjoining regions of Europe and Asia. The family also includes Antarctic Pearlwort *Colobanthus quitensis*—the southernmost dicot in the world—and one of the only two flowering plants that occur on the Antarctic continent.

Including naturalised species, there are around 100 species in the family in the UK. Some, such as Corncockle, campions, pinks and catchflies, are decorative and well known but many others, including stitchworts, chickweeds, mouse-ears, sandworts and sea-spurreys, are relatively dull, inconspicuous and often tiny. The genus *Dianthus* includes popular, ornamental garden plants, such as Carnations *Dianthus caryophyllus*, Sweet William *Dianthus barbatus* and pinks. Carnations are the most valuable commercially and especially popular as cut-flowers. They have been bred and hybridised to produce hundreds, or even thousands, of varieties in all shades of red, pink and white. Many have a spicy fragrance and are used in the making of perfumes. Several species of campions *Silene* are common in the UK and attractive to pollinators. They often find their way into gardens.

Corncockle *Agrostemma githago*

Corncockle is an ancient introduction that arrived in Britain from Europe in 'seed-corn' in the Iron Age, probably in about the 6th century BC. Much later, it was introduced to many temperate regions of the world, including much of the USA and parts of Canada, Australia and New Zealand, presumably as a contaminant in imported European wheat.

Although never particularly common, the Corncockle declined dramatically during the 20th century with the development of intensive mechanised farming, herbicides and improved seed cleaning methods. Compared with other contaminant plants, the Corncockle was particularly vulnerable because it's very large seeds were easily screened out. It is now virtually extinct in the wild, though it is often included in seed mixtures used in wildflower gardens, in newly planted meadows on motorway and roadside verges, and in city parks.

Corncockle and Marmalade Hoverfly *Episyrphus balteatus*

All parts of the Corncockle are known to be poisonous but it has long been used in folk medicine and cases of serious poisoning by any plant are very uncommon in the UK. This did not stop *The Telegraph* newspaper, in 2014, from accusing the BBC's *Countryfile* programme of spreading poison across the land, simply because viewers were offered free packets of wildflower seeds containing Corncockle seeds. Clearly, reporters at *The Telegraph* must have been ignorant of the fact that many of the most popular plants that grow in British cottage gardens are poisonous. Examples include such favourites as Laburnum *Laburnum anagyroides*, hydrangeas *Hydrangea* spp., Foxglove *Digitalis purpurea*, Monk's-hood *Aconitum napellus*, Pasqueflower *Pulsatilla vulgaris*, Larkspur *Consolida ajacis*, daffodils *Narcissus* spp., crocuses *Crocus* spp. and bluebells *Hyacinthoides* spp. Other notoriously poisonous plants found in some gardens, including our own, are Hemlock *Conium maculatum*, Deadly Nightshade *Atropa belladonna* and Henbane *Hyoscyamus niger*.

Corncockle is an annual plant that grows up to about a metre tall with long, softly hairy leaves. It flowers from about June until August. Each of the plant's branches is tipped with a single pinkish-purple, scentless flower, surrounded by five long, pointed sepals. The flowers are 25–50 mm across and each petal bears two or three discontinuous black lines (nectar guides) converging towards the nectary.

Corncockle is an attractive plant that needs our help to survive. It is not especially attractive to pollinators, at least not in our garden, but gets visits from a few hoverflies, butterflies and at least a few bees. It apparently attracts more insects elsewhere. Once pollinated, the flowers produce a flask-like, seed pod containing large, black, textured seeds.

Red Campion *Silene dioica*

Red Campion is native to most of Europe. In the UK, it is common and familiar more or less throughout the country below about 1,000 m. Red Campion is an indicator species of ancient woodland, though nowadays it is widespread almost everywhere—in light shade in woodland clearings and rides, in hedgerows, on roadside verges, in fields and meadows and, in upland areas, on rocky slopes and scree.

In traditional medicine, crushed Red Campion seeds were used to treat snakebites, which probably explains the alternative name Adders' Flower. Folklore also suggests that if you bring Red Campion into your house, then you will get bitten by a snake the next time you go out. A more realistic use of Red Campion involves the roots, which contain saponin and can be boiled up to make a useful substitute for soap.

Red Campion is a biennial or perennial herbaceous plant that grows up to 1 m tall. Its

Red Campion

bright rosy-pink flowers, 18–25 mm across, have five deeply notched petals and lack scent. The flowers are dioecious and found on separate plants. Male flowers have 10 stamens, females five styles. The flowering season is very long, from March to November and sometimes throughout the winter in Devon and Cornwall. The flowers produce plentiful nectar and attract diverse pollinators, including bees, flies and butterflies. The longish corolla tubes of Red Campion are adapted to be visited by bees and flies that have longish tongues. The Common Snout-hoverfly *Rhingia campestris*, for example, is a regular visitor.

Red Campions and White Campions *Silene latifolia*, the latter a Bronze Age introduction, often cross-pollinate and produce fertile hybrids. Hybrids between the two species have pink flowers but subsequent backcrosses can produce flowers in any shade of pink between white and red.

Polemoniaceae (Jacob's-ladder and Phlox family)

This family includes 18–25 genera and perhaps as many as 400 species, most of them found in North America and temperate parts of western South America. The centre of diversity is in western North America, particularly California. Two genera—*Phlox* and *Polemonium*—are also found in cool, temperate regions of Asia but only *Polemonium* is found in Europe. Both genera are widely distributed in North America, suggesting that they are likely to be relatively recent arrivals in the Old World. Here we are concerned only with Jacob's-ladder *Polemonium caeruleum*. Several other species of Jacob's-ladder grow at high altitude on mountains and there is one species in the Andes of South America.

The family includes a few woody shrubs or small trees but most species are herbaceous annuals or perennials. The family has very little economic importance, though some plants are very attractive and widely grown in gardens as ornamentals, notably species of *Phlox*, *Gilia* and *Polemonium*. The flowers are quite diverse in structure and colour, reflecting the fact that the different species are adapted to attract pollinators as varied as bats, hummingbirds, bees, butterflies, moths, flies and beetles.

Jacob's-ladder *Polemonium caeruleum*

Jacob's-ladder is native to temperate regions of Europe and Asia but is widely naturalised outside its natural range. In the UK, it has a restricted native range, above about 200 m, on steep limestone screes in the Peak District and Yorkshire Dales (notably at Malham Cove) and also occurs in one area of river-side cliffs in Northumberland. It also occurs down to sea level as a more widespread garden escape, usually near habitations.

Nowadays, Jacob's-ladder is hardly ever used as a medicinal plant but that was not always the case. It was first used as a medicinal herb by the ancient Greeks who used in to treat dysentery and toothache. In the 19th century, it was also used by some European pharmacies to treat rabies and as an antisyphilitic agent.

Jacob's-ladder is a hardy, erect clump-forming perennial that grows to 90 cm tall or more and owes its name to its ladder-like, pinnate leaves—an allusion to the ladder, or stairway to heaven, on which angels were ascending or descending, that was described in the dream of the biblical Patriarch Jacob (Genesis 28). Jacob's-ladder prefers moist, well-drained soil but tolerates full sun or partial shade. It flowers in early summer, beginning in June and continuing through July, sometimes later. The plant produces loose clusters of elegant, cup-shaped blue flowers, about 2–3 cm across, with prominent, contrasting yellow-orange stamens. Jacob's-ladder is very attractive to bumblebees but not much else, other than a few hoverflies and an occasional butterfly. Once pollinated, the flowers produce seeds very freely and the plant spreads easily.

Numerous cultivars of Jacob's-ladder are available in colours ranging from white to yellow, pink or various shades of lavender or purple. In our personal opinion, none is more attractive than the natural wild form.

Jacob's-ladder and Early Bumblebee *Bombus pratorum*

Primulaceae (Primrose family)

The Primulaceae are found more or less worldwide, though most species are found in temperate areas of the Northern Hemisphere, including colder Arctic and Alpine regions. The Primulaceae was formerly considered to include about 28 genera and 800–900 species though generic limits were not at all clear. However, the family (*sensu lato*) has recently been enlarged to include the formerly recognised families Myrsinaceae, Theophrastaceae and Samolaceae. It is now estimated to include about 55 genera and 2,790 species, most of them herbaceous perennials and a few annuals. Note that the Primulaceae is not related to the Evening-primrose family (Onagraceae).

In the UK, the Primulaceae includes about 30 species, including several that are naturalised. Native species include such familiar and popular flowers as the Primrose and Cowslip in the genus *Primula*, and Scarlet Pimpernel and loosestrifes in the genus *Lysimachia*. Cyclamens (or sowbreads) are popular garden or houseplants that are increasing in the wild as garden throw-outs.

Primrose *Primula vulgaris*

The Primrose's native range includes much of Europe (and a bit of North Africa), extending south from central Norway to southern Portugal and Algeria and east from the UK through southern Europe to the Crimea, Balkans, Syria, Turkey and Armenia.

The Primrose is common throughout most of the UK except for northern Scotland. It is an ancient-woodland indicator plant but occurs widely in deciduous woodland clearings, shady hedges and nowadays even on motorway verges and railway embankments, where over-collecting by people is more risky. To prevent over-collecting, the picking or digging up and removal of Primrose plants from the wild is illegal in the UK (the Wildlife and Countryside Act 1981).

Primroses bloom early, especially in mild winters, and are welcome sign of spring. They sometimes appear as early as late December and often continue into May. They are one of the first woodland flowers to appear and provide welcome nectar to early pollinators.

Primroses promote cross-fertilisation by being flowers that are heterostylus—i.e. some are 'pin' flowers with a long style and short stamens, while others are 'thrum' flowers with the relationship reversed. As a result, pollen is deposited on two distinctly separate places on the long tongue of visiting pollinators, and it follows that pollen collected from pin flowers will be accurately transferred to thrum stigmas, and vice versa (see p. 54). Regular long-tongued visitors to Primroses include bee-flies, long-tongued bees and a few butterflies.

Primrose and Dark-edged Bee-fly *Bombylius major*

Ericaceae (Heather family)

The heather and heath family has an almost worldwide distribution (excluding Antarctica) and contains over 4,000 species in about 125 genera. The majority of species are found growing in acidic soils in temperate or cool tropical regions. Ericaceous species range from small herbaceous plants to shrubs and small trees. Dwarf-shrub communities are often the dominant plants found growing on the soils characteristic of heathland, moorland and peat bogs in Europe. Members of the family, including many epiphytes, are also very abundant in the 'ericaceous belt', often found in cloud forests on tropical mountains. Because they mostly live in acidic, nutrient poor soils, many ericaceous species live in association with fungi (ericoid mycorrhizal fungi) that inhabit their roots. The fungi facilitate the uptake of nutrients, especially nitrogen.

A high proportion of ericaceous species are cultivated as ornamentals, the most noteworthy being rhododendrons, azaleas and many species of heather. Other members of the family include blueberries, cranberries and huckleberries, all of which are of economically important crops in various parts of the world.

Some ericaceous species, notably rhododendrons, have large, conspicuous flowers and most are pollinated by bumblebees or birds, especially sunbirds. In temperate regions ericaceous species often have small bell-shaped flowers that attract bees and a few species are wind-pollinated. In Neotropical cloud forests, there are many epiphytic species which have showy, tubular, red and white flowers that secrete copious nectar. They tend to be pendant and the majority attract and are pollinated by hummingbirds. Some species have capsular fruits with wind-dispersed seeds. Others, like the blueberries and cranberries already mentioned, have fruits that attract mammals or birds that swallow the fruits and excrete the seeds.

In the UK, the family includes over 20 species in a dozen or more genera, with about 15 that are native. The family is very variable in its vegetative and superficial floral characters. Rhododendrons are large shrubs, sometimes reaching heights of 5 m, and most of the heathers and heath are small shrubs. By contrast, wintergreens are small herbs and Yellow Bird's-nest *Hypopitys monotropa* is a saprophyte lacking chlorophyll. Only two species—Heather and Bell Heather—will be considered here.

Heather or Ling *Calluna vulgaris*

Heather or Ling is the only species in the genus *Calluna*. It is native and widely distributed in Europe, Iceland and Turkey and has been introduced to suitable areas in North America, Australia, New Zealand and the Falkland Islands. In the UK, Heather is common more or less throughout but scarcer in the east of England.

Heather or Ling [PC]

Heather with Heather Colletes *Colletes succinctus* [PC] and Heather Mining Bee *Andrena fuscipes* [PC]

Heather is a shrub that grows to about 80 cm tall or sometimes a metre or more. It grows on poor, acidic soils and is usually the dominant plant on moorland, heaths, bogs and to a lesser extent in open woodland with peaty or acidic soils. Heather is tolerant of grazing and regenerates quickly after fires. In fact, grouse moors and nature reserves are routinely managed by sheep or cattle grazing and by controlled burning.

In the past, Heather was used in many different ways—as fuel, fodder for animals, for making thatch and other building materials, for packing, for making ropes and brooms, and to tan leather and dye wool yellow. It was also used to brew heather-beer, before the use of hops, and is still used for the production of heather honey.

Heather flowers from about mid-July until the end of September, often turning vast areas of moorland and heath a beautiful purple in late summer. The flowers are in dense spikes and are small, four-petalled and pinky-purple. They are popular with many bumblebees (not the long-tongued species) and solitary bees. In fact, several of the latter are heather specialists, including the Heather Colletes *Colletes succinctus*, the Heather Mining Bee *Andrena fuscipes* and the Small Sandpit Mining Bee *Andrena argentata*. In more northern climes (e.g. the Faroe Islands), where bees and any other pollinators are rare, thrips are regular pollinators of Heather and some other flowers.

Heather is an important food source for upland sheep and deer which graze the growing tips of Heather. And on grouse moors, Red Grouse feed on the young shoots and seeds. The larvae of the spectacular Emperor Moth *Saturnia pavonia* also feed on Heather.

Bell Heather *Erica cinerea*
Bell Heather is a widespread native species in western and central Europe, including the Faroe Islands. In the UK, it is widespread except in the East Midlands where is mainly absent. Bell Heather is found in various habitats, including moors, heaths, coastal heaths, dune slacks and even open woodland. It does best on well-drained, acidic, nutrient poor soils but survives on chalk or limestone heaths in full sun wherever alkaline soils have been sufficiently leached to become neutral or acidic.

Bell Heather is an excellent source of nectar for pollinators. Indeed, it was rated in the top five flowers for nectar production in a survey conducted by the AgriLand project which is supported by the UK Insect Pollinators Initiative. The heather honey collected from bees that feed on Bell Heather is dark, fragrant and very popular.

Bell Heather is quite distinctive. Its dark purple-pink, bell-shaped flowers bloom between July and September or October, covering moors and heaths with a glorious pink carpet. Its

Bell Heather [PC] and Silver-studded Blue *Plebejus argus* [PE]

abundant nectar is imbibed by many insects, particularly bees. In some areas, it is also reported to be important for Ruby Tiger *Phragmatobia fuliginosa* moths and rare Silver-studded Blues *Plebejus argus*.

Hydrophyllaceae (Waterleaf and Phacelia family)

The taxonomic status of plants in the Hydrophyllaceae remains uncertain. Traditionally, they were given family rank but some more recent authorities, notably the Angiosperm Phylogeny Group, demote them to subfamilial rank within the Boraginaceae. This treatment remains uncertain and some botanists, including Stace, continue to recognise the Hydrophyllaceae.

The family includes about 20 genera, 16 of which are native to western North America, and around 300 species. In the UK, the best known family member is probably Tansy-leaved Phacelia or Scorpionweed.

Tansy-leaved Phacelia or Scorpionweed *Phacelia tanacetifolia*

Tansy-leaved Phacelia is native to the deserts of the south-western USA and north-western Mexico. It favours stony hillsides and reaches altitudes of 2,000 m above sea level. It was introduced to the UK in 1832 and was first found growing wild in 1885.

In both its native range and Europe, Tansy-leaved Phacelia is a versatile plant. It can be used in vineyards and field borders to attract pollinators, as a cover crop, as green manure, and as an ornamental. It is not frost hardy but easily overwinters when winter temperatures are mild.

Tansy-leaved Phacelia is an annual herb, reaching a height of about a metre. It has the coiled inflorescences (scorpioid cymes) typical of many boraginaceous plants. It flowers for several months, often from as early as mid-May until about September. The blue or mauve flowers have long, prominent, whiskery stamens. They produce copious nectar throughout the day, together with abundant pollen, and so attract many bumblebees and

Tansy-leaved Phacelia

Honey Bees. In fact, it has been described as *"perhaps the single most attractive plant for bees on the planet"*, though his not a conclusion with which we, personally, would agree. Tansy-leaved Phacelia also attracts hoverflies which are potentially beneficial because their larvae eat aphids and other pests.

Boraginaceae (Borage and Forget-me-not family)

Worldwide the Boraginaceae includes over 2,700 species arranged in about 150 genera. The family is particularly diverse in the Mediterranean regions of Europe and Asia and in Central and South America. Most species are annual or perennial herbs, though there are some woody lianas, shrubs and trees, mainly in the tropics. For example, Neotropical species of *Cordia*, such as *Cordia alliodora*, reach heights of 30 m and are harvested for timber.

British floras include 30+ species belonging to the Boraginaceae but this total includes ancient introductions and several garden escapes. All British plants are perennial or annual herbs and most have hairy stems and leaves. The flowers of many species, including comfreys and forget-me-nots, occur in a characteristic coiled inflorescence (a scorpioid cyme) which coils like a scorpion's tail. The lower flowers open first. Flower colour in the Boraginaceae is predominantly blue but may be pink, yellow or white in a few species. The buds and flowers of numerous species are also noteworthy for changing colour from a pinkish-red as they open to a beautiful blue or purple as they mature. Flower structure is also variable, particularly in the length of the nectar tube. The long, pendant, tubular flowers of comfreys are at one extreme while the flowers of forget-me-nots, with a short tube, are at the other.

Relatively few species of Boraginaceae are important economically. Several tree species are used for timber production. The Neotropical *Cordia alliodora*, for example, is sometimes grown in plantations. It grows rapidly and often out-competes other trees that grow slower. In fact, outside of its native range, *Cordia alliodora* is regarded as a problematic invasive species. The family also includes numerous garden ornamentals, notably heliotropes, borage and forget-me-nots. A few species have been used medicinally. Common comfrey, for example, has been used in folk medicine and has or had a reputation for being useful for mending bones. However, common comfrey is not recommended for internal use because many species of Boraginaceae are known to contain toxic alkaloids. Flour and hay contaminated by boraginaceous weeds sometimes cause serious cases of poisoning in both humans and livestock.

In the rainforests of Central and South America the toxic alkaloids found in some species of Boraginaceae (e.g. in the genera *Tournefortia* and *Heliotropium* and Asteraceae are collected and sequestered by lek-forming glasswing butterflies (Ithomiinae) and used in the production of pheromones. The pheromones are used by male glasswings to attract females. Later, when mating takes place, some alkaloids are passed to the females to render them distasteful to predators.

Viper's-bugloss *Echium vulgare*

Viper's-bugloss is a native species in much of Europe and in temperate areas of western and central Asia. It is also naturalised in parts of North America, mainly in the north-east. In the UK, it thrives in chalk grassland and on disturbed ground, including quarries, roadsides and railway lines. Around the coast it grows on cliffs, shingle and sand dunes, sometimes *en masse* covering large areas. It is so spectacularly abundant on the Marske sand dunes in Yorkshire that the area is known as 'the Blue Mountain'.

Viper's-bugloss is a gorgeous biennial growing to about a metre tall. Its stems and leaves are roughly hairy and the bright blue flowers are arranged in dense spikes. The buds are purplish-pink but quickly turn blue as they open. The flowers are 15–20 mm long, trumpet-shaped with four or five long, reddish or purplish, protruding stamens and a forked style. The pollen is blue.

For some reason, Viper's-bugloss has a serpentine or viperish reputation. It is said to have a spotted stem that resembles snakeskin, while its flowers, with their protruding stamens, supposedly look like the head of a snake about to strike. And in the past, it was recommended

Viper's-bugloss and Early Bumblebee *Bombus pratorum*

for use to treat snake bites. None of this makes much sense to us. Viper's-bugloss is an attractive plant with beautiful blue flowers—no more, no less.

Viper's-bugloss has a long flowering season, from about June to about the end of August. It produces copious nectar and attracts a spectacular diversity of visiting insects, mainly bumblebees, hoverflies and butterflies. It is said to be a particular favourite of the Painted Lady *Vanessa cardui*.

Green Alkanet *Pentaglottis sempervirens*

Green Alkanet is native to south-western Europe but was introduced to British gardens over 300 years ago and recorded growing wild by 1724. Nowadays, it is common in damp, shady woodland margins, verges and hedgerows, usually not far from habitations. It occurs throughout most of the UK but becomes scarcer in the midlands, the north and Scotland. Because of its ability to regenerate from damaged roots and self-seed, green alkanet is often regarded as a troublesome weed. On the other hand, the flowers are pretty and attractive to pollinators.

Green Alkanet is a perennial plant growing to about 60–90 cm tall. The young plant is soft and fuzzy and easy to handle. Once older, the stems and leaves become covered with prickly bristles that are painful when touched and can cause a rash. Presumably the bristles help protect the leaves from herbivorous insects and mammals. Green Alkanet blooms in spring

Green Alkanet and Green-veined White *Pieris napi*

and early summer from April to June or July. The blue flowers are a little larger than those of forget-me-nots but otherwise similar. A white ring surrounds the central entrance to the nectar tube and acts as a nectar guide. The stamens and stigma are hidden within the central opening to the nectar tube. As in several other boraginaceous plants, the buds of green alkanet are pink when they first open but quickly change to a brilliant blue.

The flowers are attractive to a diverse range of potential pollinators, including bumblebees, solitary bees, and a few hoverflies. Occasional visits from butterflies include Orange-tips *Anthocharis cardamines* and Green-veined Whites *Pieris napi*.

Borage *Borago officinalis*

Borage is a fast-growing annual herb native to the Mediterranean region including North Africa. It was introduced into Britain by the end of the 12th century and was growing in the wild by the late 18th century. Originally Borage was cultivated for culinary use. Though hairy, the leaves have a fresh taste, reminiscent of cucumber, and are used in salads, soups and garnishes. The beautiful blue, star-shaped flowers have a sweet taste and are used to add colour to salads, desserts and drinks. After being frozen into ice-cubes, the flowers are used to decorate summer fruit drinks.

In herbal medicine Borage has long been used for numerous ailments. However, it should be used with caution because, like

Borage and Patchwork Leaf-cutter Bee
Megachile centuncularis

many boraginaceous plants, it contains toxic alkaloids which can cause liver damage. Nowadays Borage is cultivated commercially to produce borage seed oil which is said to work wonders for itchy skin complaints, including eczema and acne. Borage oil is also said to be useful to treat depression, premenstrual syndrome and menopausal symptoms.

Borage grows well in the UK's climate and self-seeds easily. It grows to about 60–100 cm tall and has star-shaped flowers 20–25 mm across. The flowering season typically lasts from May or June until the first frosts in October or November but, in mild winters, flowers are likely to appear in any month, even January or February. The flowers produce abundant nectar and are buzz-pollinated. Many bees, including Honey Bees, are attracted by the nectar but only bumblebees and solitary bees, such as leaf-cutter bees *Megachile* spp. and sharp-tail bees *Coelioxys* spp., are capable of buzz-pollinating the flowers and releasing pollen. Borage is often used as a 'companion plant' in vegetable and herb gardens. It is so attractive to bees and other pollinators that rates of pollination are enhanced, resulting in higher yields. And, by attracting parasitic wasps and flies, it also contributes to the control of pests, such as aphids and caterpillars.

Wood Forget-me-not *Myosotis sylvatica*

Most species of *Myosotis* occur in western Eurasia (about 60 species) and New Zealand (about 40 species) with only a few elsewhere in North and South America, and New Guinea. It is also widely planted in parks and gardens in temperate regions of the world.

In the UK, the Wood Forget-me-not is one of half-a-dozen or more rather similar species of forget-me-not. It is widespread below 500 m, but much commoner in southern and eastern England than in the north and elsewhere. Cultivated garden varieties of this attractive flower are very popular. The precursor of the early garden varieties of forget-me-nots is thought to have been the Water Forget-me-not *Myosotis scorpioides* but times have changed and most are now cultivars of the Wood Forget-me-not. The romantic name 'forget-me-not' originates from Germany where a knight fell into a river while collecting Water Forget-me-nots for his lady. As he was swept away by the current, he threw the flowers to his lady and cried out *"Vergiss mein nicht!"* (*"forget me not!"*).

In the wild, the Wood Forget-me-not is a short-lived perennial, or sometimes biennial, found in ancient woodland, woodland rides and woodland edge, and increasingly in hedgerows as a garden escape. The Wood Forget-me-not is in flower from about April to June and its flowering stems grow to about 20–40 cm tall, with hairy leaves and pretty blue flowers about 8 mm across. The attractive, yellow ring surrounding the centre of forget-me-not flowers is a nectar guide, advertising the whereabouts of the nectary. The ring changes colour to an unattractive dingy white once the flowers have been pollinated. The most important pollinators are probably Dark-edged Bee-flies *Bombylius major*.

Wood Forget-me-not and Red-headed Cardinal Beetle *Pyrochroa serraticornis*

Solanaceae (Nightshade family)

The Solanaceae has a more or less worldwide distribution except for Antarctica, reaching its greatest diversity in Central and South America. The family is poorly represented in temperate regions. In total, it consists of about 2,700 species in about 100 genera, including trees, shrubs, lianas, epiphytes and both annual and perennial herbs.

The family is very important economically. Agricultural crops include potatoes, tomatoes, aubergines (eggplants), chilli peppers, bell peppers and Cape gooseberries. The family also includes tobacco and numerous plants (e.g. Thorn-apple *Datura stramonium*, mandrake *Mandragora*, Deadly Nightshade *Atropa belladonna* and Henbane) that are toxic and/or contain potent hallucinogenic alkaloids, including scopolamine and atropine. And popular garden plants include *Petunia*, *Browallia*, *Nicotiana*, *Brugmansia* and many others. In the tropics, where most solanaceous species occur, their flowers attract an exceptionally diverse variety of pollinators, including nectarivorous bats, hummingbirds, bees, hawk-moths and many other insects.

Only three members of the family are native to the UK—Deadly Nightshade, Bittersweet and Black Nightshade *Solanum nigrum*. In addition, Henbane and Thorn-apple are ancient introductions, recorded in the wild for centuries and numerous ornamentals, notably Apple-of-Peru, and vegetables are regular garden escapes. The fruits of many solanaceous species are very poisonous to man but it does not seem to be clear whether the toxins are concentrated in the fruit pulp, the seeds or both. The fact that birds eat many berries but excrete the seeds perhaps suggests that most of the poison is in the seeds.

Henbane *Hyoscyamus niger*

Henbane, also called Stinking Nightshade, is a highly toxic plant, native to Eurasia and naturalised throughout much of the world, including the United States and Canada. In the UK, Henbane was introduced in the Bronze Age and is now listed as Vulnerable. It is a lowland species, generally rare but sometimes locally common in the south and east of England. It

Henbane [PC] and Marmalade Hoverfly *Episyrphus balteatus*

is characteristically found on dry calcareous or sandy soils, often by the sea, associated with disturbed waste ground, or in the vicinity of rabbit warrens.

All parts of Henbane are poisonous but the plant has such a foul odour that it is seldom grazed by livestock and very few cases of livestock poisoning have been recorded. Nevertheless, it is a plant that is packed full of poisons, notably the tropane alkaloids hyoscyamine, atropine and scopolomine. Symptoms of poisoning are said to include delirium and convulsions that lead to coma and death. In the Middle Ages, Henbane was second only to Monk's-hood *Aconitum napellus* as a favourite of poison-makers (and poisoners?). In modern times, Henbane achieved fame and notoriety when Dr Crippen, an American homeopath, used it as a source of poison to murder his wife. Crippen was hanged in Pentonville Prison in 1910.

In spite of its poisonous properties, Henbane has been used medicinally since the 10th century. It has been used as a sedative and/or narcotic to treat a great variety of ailments, including travel sickness, toothache, earache, arthritis, mental illness, maniacal excitement, sleeplessness and nervous depression. It has also been valued in surgery for its useful narcotic effect. It still has limited uses today—scopolamine is an active ingredient in travel sickness patches and atropine is used by opticians in eye drops to dilate the pupil before eye examinations.

Henbane is a hairy, sticky, foul-smelling, biennial plant that grows up to 80 cm tall. The flowers are 2–3 cm across and trumpet-shaped. They are coloured a distinctive creamy-yellow, netted with purple veins, and with a dark purple throat. They are often described as evil- or sinister-looking for no very obvious reason. Henbane is in flower from June to August. Though a scarce and interesting plant, Henbane is not particularly attractive to pollinators. Hoverflies are the most frequent visitors and other flies are presumably attracted by the plant's unpleasant smell.

Apple-of-Peru or Shoo-fly Plant *Nicandra physalodes*
Apple-of-Peru is native to western South America, the centre of its range being Peru, Bolivia and northern Chile. It has been widely introduced and often naturalised in tropical and temperate countries worldwide, including the USA , China, South East Asia, Australia, East Africa and elsewhere. In the UK, it is a frequent garden escape, often found on waste ground and rubbish tips where its presence is usually due to seeds originally included in commercial bird seed mixtures.

An alternative name for Apple-of-Peru is Shoo-fly Plant which comes from its use as a fly repellent. When rubbed on exposed skin, its foliage releases a sticky, toxic substance that repels flies but can also result in hallucinations for the user. All parts of the plant are regarded as poisonous. Nevertheless, the plant is occasionally harvested from the wild for both local food and medicinal uses.

Apple-of-Peru is an annual that grows to a height of up to 80–120 cm. It makes an attractive garden plant that blooms from about June to September, bearing a succession of solitary flowers that remain open for just a few hours. The beautiful, bell-shaped flowers, about 4–5 cm across, are pale blue with a white throat. They attract a few insect pollinators, mainly bumblebees and a few butterflies, but are not especially popular. The cherry-like, green-brown berries that follow flowering are encased in a papery calyx that resembles a miniature Chinese lantern. These ornamental, globular fruits are often dried and used as winter decorations.

Apple-of-Peru

Bittersweet or Woody Nightshade *Solanum dulcamara*

Bittersweet is a vine that is native to much of Europe, Asia and North Africa and widely naturalised elsewhere in the world. In the UK it is a widespread lowland plant but absent from much of Wales and Scotland. It is a semi-woody, perennial vine, common in woodland, scrub and hedgerows, that scrambles over other vegetation to heights of 2 m or more.

Bittersweet is in flower from about June to September. The flowers are up to 1.5 cm across, in loose clusters, with five purple petals and a cone of prominent yellow stamens surrounding the style. The flowers have to be buzz-pollinated (p. 47) and attract bumblebees and a few smaller bees (but not Honey Bees that are unable to buzz-pollinate).

When ripe, the fruits are red berries that average about 8–9 mm in diameter, making them accessible to a good range of the birds that are the main seed dispersers. The first berries ripen in late July and others follow throughout August and September. They are watery but thin-skinned and soon shrivel and dry up if not eaten. The tempting berries are poisonous to humans and livestock but so bitter to taste that cases of accidental poisoning are rare. The berries are eaten by many birds, particularly Blackbirds, Song Thrushes, Robins and Blackcaps, but are not very nutritious and often ignored when preferred alternatives are available. Bullfinches are known to be persistent seed-predators and sometimes destroy a considerable proportion of the seed crop.

Bittersweet [PC] and Buff-tailed Bumblebee *Bombus terrestris*

Veronicaceae (Foxglove and Speedwell family)

Based on recent genetic studies using DNA sequencing, the Angiosperm Phylogeny Group has recommended a major enlargement of the Plantaginaceae to include several former families. The latter include the Veronicaceae and Scrophulariaceae, both of which are demoted to subfamilies. In older, traditional classifications, the Plantaginaceae contained only three genera with over 250 species, all but three of them in the genus *Plantago*—the plantains. However, Stace does not follow the APG's recommendations and continues to recognise both the Veronicaceae and Scrophulariaceae as distinct families.

Following Stace, the Veronicaceae in the UK includes 11 genera and over 50 species, many of them introductions and many formerly in the Scrophulariaceae. Familiar species that are placed in the Veronicaceae by Stace include the foxgloves, toadflaxes and speedwells.

Foxglove *Digitalis purpurea*

In the past, foxgloves were usually placed in the family Scrophulariaceae, along with figworts and mulleins, but recent research links them to snapdragons, toadflaxes and speedwells, while figworts and mulleins remain in the Scrophulariaceae.

There are about 20 foxglove species in the genus *Digitalis*. Our familiar Foxglove is native to Europe and widespread in the UK below about 900 metres. It is a plant of woodland clearings and heaths and often associated with acidic soils. It is also common in gardens and escapes have enabled it to spread well beyond its native range. It is now naturalised in many parts of the world. For example, we found it to be very common in the Talamanca mountains of Costa Rica.

The Foxglove has long been used in folk medicine to treat various ills, notably dropsy (oedema due to congestive heart failure). The Foxglove's common names include Dead Man's Bells and Bloody Bells, appropriate considering that all parts of the plant are toxic and a therapeutic dose is very close to a lethal dose. The pharmacological properties of the Foxglove were investigated by the Scot William Withering, an 18th century botanist and physician, who recorded his findings in his influential book *An Account of the Foxglove*, published in 1785. He demonstrated that, in appropriate precise doses, the main active ingredient—digitalis—is beneficial for heart conditions and is an effective diuretic.

The Foxglove is a biennial with a flowering season lasting from about June to August or September. Its unbranched flowering spikes feature up to 80 flowers and reach a height of up

Foxglove [PC] and Common Carder Bumblebee *Bombus pascuorum* with pollen on its thorax

to two metres. The pinkish-purple (sometimes white) flowers are 3–5 cm long and more or less tubular in shape. Foxglove flowers are protandrous with anthers that release pollen before the stigma is receptive. The flowers are adapted to be pollinated by bumblebees that land on the enlarged, spotted lower lip. A bumblebee is quite large and only just fits into the flower. As it squeezes in, pollen collects on its back as it brushes against the stamens on the roof of the flower.

Foxglove flowers open sequentially, starting from the bottom and each remains open for several days. Whenever several flowers are open at the same time, which is usually the case, those that are lower down will be in the female phase, the upper ones in the male phase. Visiting bumblebees habitually fly directly to the lower flowers and then move upwards. If they arrive carrying pollen from another plant, it will probably be left on the stigma of the lower flowers that are likely to be in the female phase and have receptive stigmas. Then, higher up before they depart, bumblebees collect fresh pollen from flowers that are in the male phase and then carry the fresh pollen to the lowest, probably female flowers on a nearby inflorescence. The pollen is then available to be deposited on the stigma of the next flower to be visited, resulting in cross-pollination. As well as bumblebees, Foxgloves are visited by a few other relatively large solitary bees—e.g. the Tawny Mining Bee *Andrena fulva* and several other large *Andrena* bees.

Germander Speedwell *Veronica chamaedrys*

Germander Speedwell, also called Angels' Eyes or Blue Bird's Eye, is a native of Europe and Asia west of the Urals, but is also found in North America and elsewhere in the world, as an introduced species. In the UK, it is very common below 1,000 m almost throughout, though it is rare in the Scottish islands of the Outer Hebrides, Orkneys and Shetland.

Long considered to be a good luck charm, the cheerful flowers of Germander Speedwell are believed to 'speed' you safely on your way. The plant has also been used in herbal medicine—in Austria to make tea for disorders involving the nervous system, respiratory tract and cardiovascular system; and in 18th century the UK as a popular drink (speedwell tea) and as a cure for gout. So much Germander Speedwell was collected in London that the plant was almost eradicated at that time.

Germander Speedwell is not always so well regarded. It is a low creeping plant that sometimes forms mats or patches on lawns. On our own lawn, the exquisite bright blue flowers, mixed with those violets, celandines and cinquefoils, are a delightful accompaniment to spring and early

Germander Speedwell and Green Furrow Bee *Lasioglossum morio* [PC]

summer. Sadly, it seems that they are not always welcomed by some who see them as a major problem on lawns, especially as they are resistant to weed killers. It is beyond our imagination how anyone could prefer boring grass to Angels' Eyes.

In the UK, Germander Speedwell is widely distributed in all sorts of grassy places—meadows, pastures, roadside verges, waste ground, lawns and woodland clearings. It has small, triangular, bluntly toothed leaves and grows to a height of about 20 cm or more. The small spikes of bright sky-blue flowers measure 8–11 mm across and have a white eye at their centre. They are in flower from March or April to July and provide an excellent source of nectar for small solitary bees (e.g. *Lasioglossum* and hoverflies.

Common Toadflax *Linaria vulgaris*

Common Toadflax, also called Butter-and-eggs, Bunny Mouths and many other local names, is an attractive plant that is native to most of Europe and northern Asia and has been introduced to North America. In the UK, it is common and widespread in England, Wales and southern Scotland, and is quick to colonise disturbed land—roadside banks and verges, railway lines, abandoned mines, construction sites, unused agricultural land, etc.

Research has confirmed that Common Toadflax has medical benefits that include diuretic and fever-reducing properties. As a tea made from the leaves, or an ointment from the flowers, it was long used in traditional medicine to treat a variety of ailments, including jaundice, dropsy, intestinal problems, skin diseases and piles. Remarkably, as a tea made in milk, not water, it was also used as an insecticide. The flowers are also the source of a yellow dye.

Common Toadflax is a perennial that grows to about 80 cm, sometimes more. Its yellow flowers, with their distinctive orange bulge and long straight spur, resemble snapdragon flowers and are tightly packed on a the tall spike. The flowers first open in June and continue until October or later. Because the flower's 'throat' is closed by its underlip, only strong insects, such as bumblebees, can force their way into the flower to access the nectar spur and pollinate the flower. In practice, it is mainly long-tongued bumblebees, such as the Garden Bumblebee *Bombus hortorum* and Common Carder Bumblebee *Bombus pascuorum*, that can access the nectar spur. The Hummingbird Hawk-moth *Macroglossum stellatarum* is also an occasional visitor. Self-pollination sometimes occurs but, as noted by Darwin, cross-pollination results in many more seeds.

Common Toadflax and Garden Bumblebee *Bombus hortorum*

Scrophulariaceae (Figwort, Mullein and Buddleia family)

In the past, the Scrophulariaceae was considered to include about 275 genera and over 5,000 species. However, recent genetic studies by the Angiosperm Phylogeny Group, using DNA sequencing, have shown that the traditional classification of the family is grossly polyphyletic, meaning that not all of the genera included are closely related in evolutionary terms. As a result, the family has been reduced drastically with numerous genera being transferred to other families. The Scrophulariaceae now includes just 62 genera and around 1,700 species. It has a more or less worldwide distribution, though the majority of species are found in temperate areas, including at high altitudes on tropical mountains. The family contains no crop plants of much economic importance but is notable for many ornamental garden plants

Following Stace, in the UK the family in its reduced form includes just 3 genera and 16 species, including garden escapes. The genera and species involved include the figworts, mulleins and buddleia.

Buddleia or Butterfly-bush *Buddleja davidii*

Buddleia is a shrub up to 5 m or more tall and native to central China. It has become an extremely popular ornamental that has been widely introduced, successfully to Kew Gardens in 1896, and later to North America, Australia, New Zealand and parts of Africa. However, buddleia is an opportunistic plant that has been classified as an invasive species in many countries and as a noxious weed in parts of the United States.

In the UK, Buddleia is naturalised and quickly colonises waste ground and even finds footholds on derelict buildings. It has become common along railway lines, its seeds being carried for miles by the slipstream of trains. Given the opportunity, Buddleia forms dense thickets liable to outcompete and eliminate native plants, including many that are important for wildlife and overall species diversity. Buddleia spreads by means of its tiny winged seeds that are produced in enormous numbers. It has been estimated that a single flower spike can produce 40,000 seeds and a mature shrub as many as three million. The charity Butterfly Conservation recommends planting Buddleia to attract pollinators but warns that it needs to be managed and suggests that gardeners remove seed heads. There is plenty of time and opportunity to do so because the seeds are produced in autumn but do not ripen and disperse until the following spring. Also, non-invasive Buddleia cultivars are available, including several that are sterile.

Buddleia with Peacock *Aglais io* and Comma *Polygonia c-album*

Buddleia with Furry Dronefly *Eristalis intricarius* and Hummingbird Hawk-moth *Macroglossum stellatarum* [PC]

Buddleia flowers for most of July and August with dense, showy spikes of tiny flowers with four petals. Flowering is asynchronous, individual flowers opening outwards along the spike from the base. Each flower lasts for one to three days and the spike for up to two weeks. The short, tubular flowers have a fragrant, slightly fruity smell and produce abundant nectar. They attract a spectacular number and diversity of insects. As the bush's alternate name suggests, butterflies are among the most conspicuous and frequent visitors. Most butterfly species that visit our garden take advantageous of Buddleia, though only rarely in the case of blues and skippers. Buddleia also attracts Honey Bees, bumblebees and many hoverflies. And it is a favourite of Hummingbird Hawk-moths *Macroglossum stellatarum*.

Common Figwort *Scrophularia nodosa*

Common Figwort is a member of a genus of about 200 species found throughout the Northern Hemisphere. Most are found in Asia with just a few in Europe and North America. Common Figwort is the most widely distributed of the three species that are native to the UK. Two others have long been naturalised. Common Figwort is present throughout most of the UK, below about 500 m, mostly in damp woodland and hedgerows or alongside ponds and ditches.

The scientific generic name, *Scrophularia*, comes from the plant's traditional use as a remedy for scrofula, a tuberculous infection of swollen lymph nodes in the neck. In fact, in the Middle Ages, Common Figwort was

Common Figwort and Early Bumblebee *Bombus pratorum*

considered to be an appropriate medicinal plant to treat almost any swellings and tumours. This supposition was based on the 'doctrine of signatures' a philosophy that was widely believed from the time of the Roman physician and philosopher—Dioscoride (130–210 AD). The doctrine states that any herb that resembles any body part can be used to treat any problem afflicting it. The bulbous shape of the rhizomes of common figwort supposedly resembles swollen glands

Common Figwort is a herbaceous perennial that reaches a metre or more tall with a distinctive square stem. The flowers are rather small, almost globular, in small clusters. Their colour is rather variable—greenish to reddish-brown. Each flower has two lips, the lower one with three lobes, the upper one with two. There are four stamens and staminode in each flower.

The flowers have a foetid smell. Flowering takes place from June until August or September. In appearance, figworts are classic examples of wasp-flowers though, in our experience, they usually attract many more visits from bees, particularly bumblebees, than they do from wasps.

Great Mullein *Verbascum thapsus*

Great Mullein is the commonest of the four native mulleins in the UK and there are several other naturalised species and garden escapes. Great Mullein is a native of Europe, Asia and northern Africa and has been introduced to North America and Australia. In the UK, it is common and widespread in England and Wales but scarcer in Scotland. It is frequently found in open scrub, dry meadows, roadside verges, railway embankments, waste ground and abandoned quarries.

Great Mullein is a biennial that spends its first year as a rosette of furry, grey-green leaves. In its second year, it grows to 2 m or more in height with numerous flowers densely clustered around its stately flowering spike. Each flower is open for only one day, opening before dawn and closing in the afternoon, but the flowering stem often continues for several months. The usual flowering season lasts from about June to August or even September if the weather is very mild.

The individual flowers are 15–30 mm across and have five stamens, though not all of them are the same. The three upper stamens are deceptive, being covered with fluffy hairs that make them more conspicuous, and attractive, to potential pollinators. The two lower stamens are functional and lack the fluffy hairs. The flowers are protogynous (the stigma is receptive before the anthers are mature and ready to shed pollen). They attract solitary bees and hoverflies but will self-pollinate if they have not already been pollinated by a visiting insect.

Great Mullein is a prolific producer of seeds. Each plant produces hundreds of capsules, each containing hundreds of seeds. Total seed production per plant can amount to 180,000 or more seeds. The seeds may remain dormant in the seed bank for many years and germinate up to 100 years later.

Great Mullein [PC] and Glass-winged Syrphus *Syrphus vitripennis*

Lamiaceae (Mint, Dead-nettle and Lavender family)

The family Lamiaceae (formerly the Labiatae) has a more or less worldwide distribution, excluding Antarctica. The Lamiaceae includes about 236 genera and over 7,000 species. Most are herbaceous but a few are shrubs, vines or even trees (e.g. Teak).The family is particularly species-rich around the Mediterranean, in central Asia, and Central and South America.

The Lamiaceae includes about 60 species in the UK, including ancient introductions and garden escapes. Most herbaceous members of the family are characterised by square stems and opposite leaves. The flowers are bilaterally symmetrical and usually bisexual. The five united sepals are fused into a calyx tube and the five petals into a corolla which is split into upper and lower lips.

Species of Lamiaceae have aromatic foliage and produce many secondary compounds, including volatile oils which can be extracted when the foliage and flowers are crushed or distilled. A few are grown commercially on a large scale. Lavender oil is distilled from various species of lavender and is a prized ingredient in the perfumery trade. Varieties of wild mint are used to produce peppermint oil, used for flavouring, as well as menthol, used in toothpastes, cough drops, beverages and tobacco. Many species are familiar kitchen herbs, including rosemary, oregano, basil, mint, thyme, sage and lavender. And some species are widely used as ornamentals, notably Lavender, Wild Marjoram, *Salvia* and *Coleus*.

The Lamiaceae includes plants that are very attractive to pollinators, including many butterflies and moths, as well as bees and hoverflies. Lavender and Wild Marjoram are two species that are exceptionally popular and continue flowering for a long time, often four to six months, or even more. Another useful plant is the hybrid cultivar *Caryopteris* × *clandonensis* with parents native to China, Mongolia, Korea and Japan. It is particularly useful because it is late-flowering, beginning in July or August and continuing until late November. It is popular with bumblebees and butterflies.

Garden Cat-mint *Nepeta* × *faassenii*

Garden Cat-mint is a perennial cultivated for its fragrant grey-green foliage and masses of violet-blue flowers. It is a hybrid variety whose parents are *Nepeta racemosa*, from the Caucasus, Turkey and northern Iran, and *Nepeta nepetella*, from southern Europe and north-west Africa. It was first cultivated in the Netherlands and has since won the Royal Horticultural Society's Award of Garden Merit.

Garden Cat-mint is a good groundcover plant that forms dense clumps of foliage up to 60 cm tall and wide. It is hardy and tolerant of drought and most growing conditions, including dry or moist soils and full sun or partial shade. It is also long-lived and resists most pests. However, it is said to be very attractive to cats that love to roll about in it and sometimes flatten it.

Its flowering season is very long, lasting from mid-May until late October or even November. The flowers are particularly attractive to Early Bumblebees *Bombus pratorum* and Common Carder Bumblebees *Bombus pascuorum*, both long-tongued species that are the main pollinators. Visitors also include butterflies, particularly Green-veined *Pieris napi* and Small Whites *Pieris rapae*, Painted Ladies *Vanessa cardui* and occasional other species.

Garden Cat-mint with Common Carder Bumblebee *Bombus pascuorum* and Green-veined White *Pieris napi*

Balm or Lemon Balm *Melissa officinalis*

Balm (or Lemon Balm) is a perennial plant that is native to south-central Europe, the Mediterranean region, Iran and central Asia. It was introduced to Spain in the 7th century and reached Britain by late in the 10th century. Thereafter, it was taken as a medicinal plant to North America by the early European colonists. In the UK, it is commonly cultivated in gardens and is now a long-established garden escape in the lowlands, mainly in southern areas.

Balm and White-footed Furrow Bee
Lasioglossum leucopus

There is evidence that Balm has been used medicinally for hundreds of years in folk remedies as a treatment for disorders of the liver, nervous system and digestive tract. In the 14th century, Carmelite nuns used Balm to prepare an alcoholic tonic that they had the nerve to call Carmelite 'water'. Nowadays, Balm is mainly used as a lemon-scented herb in herbal teas or as flavouring for ice cream, candy, fish dishes and even toothpaste. When crushed, the leaves of Balm can be used to extract an expensive, essential oil for use in perfumery and aromatherapy. The oil is particularly popular in aromatherapy where it is believed to combat difficulty in sleeping, to promote calmness and to ease stress.

Outside of gardens, escaped Balm is found in hedgerows, on roadside verges, in scrub and on waste ground. Balm grows to a height of about 70 cm, sometimes higher. The small white flowers, with a corolla only 8–15 mm long, are rather inconspicuous. They bloom from late July to September and produce plentiful nectar that attracts numerous bees, including Honey Bees. In our garden, the flowers attract only small bees, especially the smaller species of furrow bees *Halictus* and *Lasioglossum* and mining bees, probably because other, more popular flowers, notably Lavender and Wild Marjoram, are in bloom at the same time.

Wild Marjoram or Oregano *Origanum vulgare*

Wild Marjoram is native to western Eurasia, including the Mediterranean region. It is a sun-loving, herbaceous perennial, about 30–60 cm tall, characteristic of dry, calcium-rich grasslands, roadsides and scrub. It is vulnerable to grazing and does not survive well on pasture but grows well in most well-drained garden soils. Wild Marjoram is the same species as the classic Mediterranean herb known as Oregano—one of the most popular of all culinary herbs and much used in pizzas. The native British plant is slightly less aromatic with a less intense flavour.

Wild Marjoram [PC]

Wild Marjoram with Hornet Hoverfly *Volucella zonaria* [PC] and Small Skipper *Thymelicus sylvestris*

Wild Marjoram has dense, rounded clusters of purple buds which open to reveal pale purplish or pinkish white flowers that secrete copious, readily available nectar. The flowering season is from June to September. Scientists at the University of Sussex counted the insects that visited 32 popular garden flowers and found that Wild Marjoram was the best all-rounder, popular with a great variety of insects, especially bees, hoverflies and butterflies.

English Lavender *Lavandula angustifolia*

Lavandula angustifolia, often called English Lavender, is a native of the Mediterranean region. It is a very common ornamental plant with many cultivars and is a widely used as an aromatic herb in cooking. It can be used like Rosemary and also adds an interesting flavour to cakes and desserts. Commercially, it is used for the extraction of lavender oil—an essential oil widely used in perfumes, aromatherapy and as a natural mosquito and moth repellent.

Lavender is an aromatic, evergreen shrub that grows to heights of up to two metres. It does best in well-drained soils in full sun and is well able to survive dry or drought conditions. It is a classic plant of the quintessential English cottage garden—much loved for its fragrant and attractive flowers

Lavender is in flower from about June to September. The pinkish-purple (lavender) flowers grow on spikes 2–8 cm long. Even though it is not native, Lavender is one of the very best plants for pollinators, attracting hordes of bumlebees, and butterflies. It is also a great favourite of migrant Hummingbird Hawk-moths *Macroglossum stellatarum* and Silver Y *Autographa gamma*

English Lavender with Honey Bee *Apis mellifera* and Wall *Lasiommata megera*

moths. It does not attract many flies. According to scientists at the University of Sussex, the Lavender varieties most attractive to bumblebees and other pollinators are Grosso, Hidcote Giant and Gros Blue.

Rosemary *Salvia rosmarinus*

Previously known as *Rosmarinus officinalis*, Rosemary is now included in *Salvia*, a genus that includes hundreds of species. Rosemary is native to the Mediterranean region and parts of Asia but is now widely grown as an aromatic herb and naturalised in most parts of the world, including North and South America. Rosemary has a long history. It was mentioned on Egyptian stone tablets as early as 5,000 BC. Pliny the Elder (23–79 AD), the Roman naturalist and philosopher, mentions it in his *The Natural History*. And Rosemary was naturalised in China as early as 220 AD during

Rosemary and Tree Bumblebee
Bombus hypnorum

the Han Dynasty. It was probably the Romans who introduced Rosemary to Britain and it was definitely naturalised by the mid 14th century.

Rosemary is best known as a fragrant evergreen herb with leaves that are used as a culinary flavouring in foods, particularly roast lamb and other roasts, casseroles and fish. But take care that its strong taste does not overpower more delicate flavours. Rosemary is also a rich source of antioxidants and anti-inflammatory compounds and has been esteemed since ancient times for its medicinal properties. It is said to alleviate muscle pain; boost the immune system and improve blood circulation; improve memory; and even promote hair growth.

Rosemary is a woody, evergreen shrub that can reach heights of 2 m or more. It thrives in well-drained soil in a sunny spot, is quite hardy in a cool climate and is drought-tolerant. Its flowering seasonality is quite unusual. It can bloom more or less constantly in a warm climate but follows a strange pattern in our Norfolk garden. There are usually a few flowers in mid-winter, followed by profuse flowering from February through May. A cessation of flowering in mid-summer, from June through September, is followed by another peak in October and November. The flowers attract plenty of bees but only bees—Honey Bees in large numbers, Hairy-footed Flower Bees *Anthophora plumipes*, a few bumblebees and a few of the larger mining bees *Andrena*. It is an important resource for any bees that emerge early in the year.

Wild Clary *Salvia verbenaca*

Wild Clary is found around the Mediterranean, including southern Europe, North Africa, the Near East and the Caucasus. In the UK, it is a locally common native in the south and south-west of England and in East Anglia but rare elsewhere. It grows mainly in chalky soils, favouring dry, sunny grassland, roadside verges, churchyards and dunes.

Wild Clary is a perennial herb, up to 80 cm tall, with hairy stems and wrinkly, grey-green foliage. Like many members of the family Lamiaceae, Wild Clary has aromatic foliage that is used to prepare a herbal tea, said to be

Wild Clary and Garden Bumblebee
Bombus hortorum

good for digestion. The young leaves can also be eaten fried or candied and the flowers can be added to salads.

In common with other claries, Wild Clary produces erect spikes of flowers arranged in whorls. The flowers are violet-blue, about 10–15 mm long, with a hooded upper lip and a three-lobed lower lip. Wild Clary flowers from May or June until about September. The flowers are bisexual (and some self-pollinate) and are very attractive to some bumblebees, particularly long-tongued species like the Garden Bumblebee *Bombus hortorum* and Common Carder Bumblebee *Bombus pascuorum*. Painted Ladies *Vanessa cardui* and a few other butterflies are occasional visitors but the structure of claries makes access difficult for many pollinators.

Asteraceae (Daisy family)

The Asteraceae has a worldwide distribution except for mainland Antarctica. With well over 1,500 genera and between 20,000 and 30,000 species, it is probably the largest plant family (rivalled only by the orchids) and accounts for as much as 10% of the native flora in most regions. The Asteraceae is an enormously diverse family, commonest in open and disturbed habitats but also present in woodland and forest. The majority of species are herbaceous but growth forms range from tiny daisies to trees 30 m tall, with a significant number of shrubs, climbers, epiphytes and aquatics in between. In the UK there are about 100 species that are native or ancient introductions and well over 20 introductions that have become established in more recent times.

Considering its size, the family Asteraceae is not especially important economically. Artichokes and lettuce are useful vegetables; sunflower and safflower seeds are crops used to make cooking oils; and various species are used to make coffee substitutes and herbal teas. The Dalmatian Insect-flower or Dalmatian Chrysanthemum *Tanacetum cinerariifolium* is the source of pyrethrum, a natural insecticide. It is curious that some Neotropical marigolds *Tagetes* are used as both a tea and a pesticide. Wormwood *Artemisia absinthium* is used as an ingredient in the spirit absinthe and as a flavouring in vermouth and bitters. Many species are used in herbal remedies and folk medicine (see species accounts). The family is perhaps most important economically as a source of popular garden ornamentals, such as chrysanthemums, dahlias, zinnias, cosmos, pot marigolds and coneflowers. On the other hand, the family is also a source of invasive species, notably Dandelion *Taraxicum officinale* agg. and Common Ragwort *Jacobaea vulgaris*, usually considered to be noxious weeds.

Blue Globe-thistle *Echinops bannaticus*

The Blue Globe-thistle is a native of south-eastern Europe, including Romania, Hungary and Serbia. In the UK, it is common garden escape or throw-out on roadsides and waste ground.

This attractive Globe-thistle is a hardy, clump-forming perennial with prickly, grey-green foliage that grows to about 1.2 m tall. It is best grown in well-drained soil in full or dappled shade. The plant dies back each autumn to below ground level but fresh growth reappears in the following spring. Its spherical, flower-heads covered with blue flowers are up to 5 cm across and bloom from July until August or September. The flowers are highly attractive to numerous butterfly species, especially nymphalids, and some bees.

Blue Globe-thistle and Median Wasp
Dolichovespula media

Spear Thistle *Cirsium vulgare*

The Spear Thistle is native to Europe, much of Asia and the Atlas Mountains of north-western Africa; and it is naturalised in East Africa, North America and Australia. It is common more or less throughout the UK mainly in disturbed areas, including overgrazed pastures, cultivated or waste ground, field edges and roadside verges. It is the commonest large thistle in the UK and the most likely model for the Scottish national flower. It is one of nine species of *Cirsium* in the UK.

The Spear Thistle has been labelled an injurious weed and is one of five species listed in the Weeds Act 1959 that requires landowners to prevent the weeds from spreading. On the other hand, the spear thistle is beneficial in that it produces lots of nectar. It was rated sixth of the ten best plant species for nectar production in a survey carried out by the AgriLand project funded by the UK Insect Pollinators Initiative. Another thistle—the Marsh Thistle *Cirsium palustre*—tops the list and most other thistles are good nectar producers and therefore good for pollinators.

The Spear Thistle is a biennial, an imposing plant up to 150 cm tall. Its stems have spiny wings and its leaves have twisted, spear-shaped lobes, tipped with impressive, sharp spines. The flower-head is composed only of tubular florets, all of which are reddish-purple and bisexual. The many florets demand many visits by pollinators for all to be pollinated. This is not a problem— the Spear Thistle's flowering season continues from about July to September and coincides with the months when bees are most abundant. All thistles are particularly attractive to bumblebees, Honey Bees and larger solitary bees but also attract many butterflies and a few flies.

Spear Thistle [PC] and Cliff Mining Bee *Andrena thoracica*

The seed production of Spear Thistles (and many other thistles) runs to 100 or more per flower-head and up to several thousand per plant. The seeds are equipped with long, feathery 'parachute' hairs and are dispersed by wind. They are eaten by Goldfinches, Linnets and various other finches.

Greater Knapweed *Centaurea scabiosa*

Greater Knapweed is a European native perennial and common in much of lowland Britain, though scarcer in Wales and Scotland. As its specific name suggests, its leaves were used in the past to treat scabies, a contagious skin infection caused by tiny mites *Sarcoptes scabiei* that burrow into the skin and cause intense itching.

Greater Knapweed is a vigorous, thistle-like plant, growing up to a metre or more tall. It is characteristic of rough grassland, verges and waste ground, usually on dry, calacareous soils.

Greater Knapweed [PC] and Marbled White *Melanargia galathea* [PC]

Unlike some knapweeds, the large pinkish-purple, composite flowers of Greater Knapweed, 30–60 mm across, consist of both disc and ray florets. The inner disc florets are fertile while the enlarged, ray-like florets are sterile and purely decorative, designed to attract pollinators.

Greater Knapweed flowers from June or July until September, providing a valuable source of nectar for much of the summer, continuing later than most other flowers. It is very attractive to insects, especially bumblebees, leaf-cutter *Megachile* and sharp-tail bees *Coelioxys*, some hoverflies and butterflies, notably browns, blues and skippers. It is said to be a huge favourite of the Marbled White *Melanargia galathea*—a butterfly that is extending its range but does not yet occur in our part of Norfolk.

Cornflower *Centaurea cyanus* and Perennial Cornflower *Centaurea montana*

The Cornflower is one of several 'cornfield weeds' introduced to Britain with the spread of agriculture across Europe during the Iron Age (from c. 1,000 BC until AD 100). It is a beautiful plant, evocative of our farming past, and once common in 'cornfields', growing in colourful abandon along with poppies, Corncockles *Agrostemma githago*, Corn Marigolds *Glebionis segetum* and other flowers. It is now endangered in the wild, driven to near extinction by modern farming methods and the use of herbicides. However, it is a popular ornamental, common in gardens, and a frequent garden escape. It has also been widely naturalised in North America and elsewhere.

Cornflower and Perennial Cornflower [PC]

The Cornflower is an annual that grows to about 50–90 cm tall and flowers from about June to August or September. The flowers are a composite head, 15–30 mm across, consisting of a central cluster of fertile, bisexual florets surrounded by a ring of much larger, ray-like florets. They are an intense blue and a rich source of nectar for the bumblebees and butterflies that are its main visitors and pollinators.

The closely related Perennial Cornflower (also called Mountain Cornflower or Mountain Bluet) is common and widespread in the mountain ranges of southern Europe but is rarer further north. It is a plant of meadows in the upper montane and sub-alpine zones. It is also a popular ornamental plant and garden escapes are well established in the wild in the UK, Scandinavia and North America.

The Perennial Cornflower is distinguished from the cornflower by being a perennial rather than an annual; and by having outer ray-florets that are longer and more deeply cut than the inner florets. The Perennial Cornflower reaches 30–70 mm tall and the flowers, 60–80 mm across, are much larger than those of the cornflower. They have pinkish-purple disc florets with long, dark blue stamens that curve inwards. The flowers bloom from May till August and attract numerous pollinators, particularly small solitary bees.

Perennial Sowthistle *Sonchus arvensis*

The Perennial Sowthistle is native to most of Europe and is considered to be an invasive noxious weed in North America, Russia, Australia, New Zealand and elsewhere. In the UK, it occurs almost throughout below 500 m, growing on arable field margins, waste-ground, roadside verges, river banks, by streams and ditches, and on the seashore, dunes and shingle. It also occurs in crops and sometimes causes substantial losses to crop yields.

Perennial Sowthistle is a medium to tall plant, in the range 60–150 cm tall, with branched stems and leaves clasping the stem with rounded, ear-like lobes, and loose clusters of showy, golden-yellow flowers, 40–50 mm across. The bracts are covered with sticky, yellowish glandular hairs. The related Prickly Sowthistle *Sonchus asper* and Smooth

Perennial Sowthistle and Patchwork Leaf-cutter Bee *Megachile centuncularis*

Sowthistle *Sonchus oleraceus* are similar but with smaller, paler yellow flowers. And Prickly Sowthistle has prickly leaves. All three of these sowthistles (sometimes called milk-thistles) have stems that bleed a milky sap which was believed, in the past, to help lactation in mothering sows. Hence the name sowthistle. All three species provide useful forage for many animals, both wild and domestic (including pet rabbits). All three species also have leaves which have been used traditionally for their anti-inflammatory properties and a root-tea to alleviate asthma and coughs.

Perennial Sowthistle flowers from late July until October and attract a reasonable selection of pollinators, including various bees, hoverflies and a few butterflies. The flowers are said to be largely self-sterile so seeds produced by self-pollination are rarely viable. Usually, around 13,000 wind-dispersed seeds are produced per plant.

Dandelion *Taraxacum officinale* agg. and lookalikes

The Dandelion is one of our most familiar flowers. But be aware that there are many confusing dandelion lookalikes, e.g. hawkbits *Leontodon*, hawkweeds *Hieracium* and hawk's-beards *Crepis*, most of which are equally attractive to various solitary bees and hoverflies.

Dandelions [PC]

The British Dandelion is native to Eurasia but similar species are found throughout much of the world. Our Dandelion is an aggregate of over 200 microspecies, able to produce seeds from unfertilised ovules, a process known as apomixis. The process produces clones which are genetically identical to the parent. Precise identification of the microspecies is best left to specialists. Examples of apomixis are also found in Dandelion relatives, notably hawkweeds, as well as in at least 33 other families of flowering plants. In an interesting development, recent research has shown that some hawkweeds have pollen that, when carried to the flowers of other nearby plants, releases toxins that kill any developing ovules, preventing them from producing seeds that would, in due course, produce plants that would compete with the hawkweed's own cloned offspring.

Dandelions are common throughout the UK, mainly below an altitude of 1,000 m, on almost any grassy area, including pastures, roadsides, wasteland and lawns. Dandelion leaves are rich in vitamins, minerals and other medicinal benefits and have been used as food or traditional medicines throughout recorded history—for thousands of years by the ancient Egyptians and Chinese. Because they have medicinal properties, dandelions were on board the Mayflower when it set sail for North America in 1620. Nowadays in the UK, the leaves are used in salads, the flowers to make dandelion wine, and the ground, roasted roots as caffeine-free coffee. In Europe Dandelions have long been used to treat liver problems and as a strong diuretic. The latter is the basis for its common vernacular names—'wet-the-bed' and 'pissy-bed'.

The Dandelion is a perennial plant. A few flowers appear throughout the year but peak flowering occurs from March until June or July. The flowers are composite heads made up of

Dandelion and
Buffish Mining Bee *Andrena nigroaenea*

numerous ray florets, each of which is a single flower. The flowers are an important source of nectar early in the year and attract a good variety of pollinators, including bumblebees, solitary bees, hoverflies and a few butterflies. As mentioned above, normal sexual reproduction, involving pollination, occurs only rarely in Dandelions. However, it is thought that pollination and fertilisation of at least a single ovule may be necessary to stimulate other ovules to produce seeds.

The flowers are quickly followed by the characteristic fluffy seed heads, each containing up to 400 seeds attached to downy white parachutes that can be blown great distances by a gentle breeze. The seeds are a useful source of food for finches, especially Goldfinches.

Fox-and-cubs or Orange Hawkweed *Pilosella aurantiaca*

Fox-and-cubs is a native of alpine regions of central and southern Europe but is now naturalised in many other temperate regions of the world, including parts of western Asia, Australia, New Zealand and North America. It was introduced to Britain in about 1620 and, being a profuse self-seeder, it quickly proliferated and became naturalised. In many parts of the world, it is now considered to be an invasive, noxious weed, the bane of many farmers and gardeners.

Nevertheless, Fox-and-cubs is an attractive flower, and a favourite of ours, that is often seen in churchyards, and on roadside verges and wasteland. It thrives in full sun. As well as being a prolific seeder, Fox-and-cubs spreads by means of creeping, horizontal runners (stolons) that take root and form new plants, often giving rise to colourful patches of flowers.

Fox-and-cubs is a perennial with leaves in a basal rosette that grows to between 30 and 60 cm in height, with flowers clustered on a slender stem. The blackish hairs that cover its stems and buds are a useful identification feature. Flowering begins in June and continues throughout the summer until October. The dandelion-like flowers are about 15 to 20 mm across and vary in colour from yellow to brilliant orange, almost red. However, the flowers reflect ultraviolet light which makes them conspicuous to pollinating insects, even those that do not see well at the red end of the spectrum. Common visitors include many species of bees, including Common Carder Bumblebees *Bombus pascuorum*, mason bees *Osmia*, numerous hoverflies and occasional butterflies. The fluffy, wind-dispersed seeds are arranged in spherical seed-heads resembling the 'clocks' of Dandelions.

Fox-and-cubs with a male Blue Mason Bee *Osmia caerulescens* and
Marmalade Hoverfly *Episyrphus balteatus*

Canadian Goldenrod *Solidago canadensis*

Canadian Goldenrod is the commonest of several non-native species of *Solidago* now found in the UK. A native of North America, it has been naturalised since the late 19th century and widespread since the 1930s. It colonises disturbed ground, roadsides and river banks and often forms dense stands by cloning. It is not shade tolerant and does not persist once shrubs and trees are established. Nevertheless, it is considered to be a serious invasive weed in parts of Europe,

Canadian Goldenrod with Yellow Dungfly *Scathophaga stercoraria* and Flesh Fly *Sarcophaga carnaria*

China and Japan. In Fukushima, it is said to have invaded and thrived in the rice fields that were abandoned after the nuclear power plant disaster of 2011.

In its native Canada, it is traditionally regarded as bringing good luck if it is found growing near to your front door. It also has other uses. If used as a dye, it produces a golden yellow colour. And tea made from its leaves is said to staunch blood loss from wounds and relieve both sore throats and urinary tract infections.

Canadian Goldenrod is now common in lowland England, but less common, scarce or absent elsewhere. It is a perennial and grows to a height of two metres or more. The numerous composite flowers are small, only 4–6 mm across, arranged on densely flowered, arching spikes. The flowers produce plentiful nectar and pollen and, in spite of their small size, are said to attract bees, wasps, flies, beetles and butterflies. In our garden, a variety of flies are by far the commonest visitors, and include hoverflies, greenbottles, flesh flies and dungflies.

Feverfew *Tanacetum parthenium*
Feverfew is native to the Balkan Peninsula, Anatolia and the Caucasus but has been widely cultivated in much of the rest of Europe, North America and Chile. In the UK, it is an ancient introduction, cultivated since AD 995. Nowadays, it is found as a garden escape throughout much of the UK.

Feverfew and Small Spotty-eyed Dronefly *Eristalinus sepulchralis*

Feverfew has been cultivated by herbalists from Europe to Egypt as a healing herb for centuries. Its earliest use is unrecorded but it has long been used as an anti-inflammatory, for fevers, headaches, arthritis and even to expel worms. The active ingredient—parthenolide—is being actively investigated and developed for various pharmaceutical applications. Nowadays, Feverfew is considered to be a useful, healing herb but one that should be used with caution. It relaxes blood vessels and is liable to interact with blood thinning medication, increasing the risk of excessive bleeding. It should be avoided by people taking blood thinners.

In the UK, Feverfew is a fairly common garden escape or throw-out, generally growing close to houses at the base of walls, on verges or waste ground. It is a herbaceous perennial, 50–70 cm tall, with pungently aromatic foliage.

The white, daisy-like flowers, about 20 mm across, are arranged in open clusters. They attract only a limited range of pollinators, mainly droneflies *Eristalis* spp. and other hoverflies or an occasional butterfly, most often a Painted Lady *Vanessa cardui* or Red Admiral *Vanessa atalanta*. Feverfew seems not to be bee-friendly. Perhaps bees are repelled by the strong, citrus scent of the flowers.

Yarrow or Milfoil *Achillea millefolium*

Yarrow is a common native in the temperate regions of Eurasia and North America. In addition, it has been introduced to Australia and New Zealand as feed for livestock. Yarrow is an interesting, aromatic plant, long used by medieval herbalists around the world. The genus was named after Achilles, the Greek warrior who is said to have used Yarrow to staunch wounds suffered during the battle for Troy. In fact, Yarrow is widely regarded as a useful, versatile herb in many cultures— a sovereign remedy for diverse ailments, including colds, toothache, hay fever, diarrhoea and many others. Being drought-tolerant and attractive to many insects, Yarrow is also widely used as a companion plant in wildlife gardens. There are many colourful cultivars—mauve, yellow, orange, pink and bright red—though, personally, we prefer to stick with the native, white version.

Yarrow is an abundant perennial throughout the UK, found in most open, grassy habitats up to altitudes of well over 1,000 m. It grows up to about 60 cm tall and has a flat-topped inflorescence with numerous flower-heads 4–6 mm across, each with five whitish ray florets and creamy disc florets. Yarrow has a long flowering season from June to October, sometimes earlier, sometimes much later. It is superficially reminiscent of an umbellifer and attracts a similar range of small, generalised insects. Frequent visitors include hoverflies, flesh flies, tachinid flies and small beetles. Visiting butterflies include an occasional Peacock *Aglais io* or Red Admiral *Vanessa atalanta* but small species, such as skippers (Hesperiidae) and Common Blues *Polyommatus icarus*, are much more frequent.

Yarrow with Varied Carpet Beetle *Anthrenus verbasci* and a hoverfly *Syritta pipiens*

Corn Marigold *Glebionis segetum*

The Corn Marigold is a native of the eastern Mediterranean but became widely naturalised in western and northern Europe with the spread of agriculture during the Iron Age. In the past it was seen as a noxious, invasive weed among crops. In Scotland, in the 13th century, a law of Alexander II decreed that any farmer letting even a single plant set seed would be fined a sheep! Along with other 'cornfield weeds', the Corn Marigold was later driven to near-extinction by modern weedkillers but is now making a comeback in agricultural set-asides and field borders and is often common in the meadows being planted in many urban and suburban parks and gardens.

Corn Marigolds in a wildflower meadow

The Corn Marigold is an annual growing to about 60 cm high. Its brilliant yellow, composite head is 35–65 mm across, consisting of female ray florets and bisexual disc florets. There is a long flowering season from June to September or even October. The flowers attract a good variety of insects, especially solitary bees, hoverflies and butterflies.

The flowers produce a prodigious number of seeds—an average of about 170 per flower head and as many as 13,500 have been counted on a single plant in a season. Most seeds fall close to the parent plant and some remain dormant and viable for several years.

Oxeye Daisy *Leucanthemum vulgare*

The Oxeye Daisy or Moon Daisy is a widespread native of Europe, including the UK, and parts of temperate Asia, including Turkey and Georgia. It has also been introduced to the USA, Canada, Australia, New Zealand and many other countries. As a garden escape the species is now widely naturalised and is listed as an invasive plant in more than 40 countries. It spreads by seeds and creeping rhizomes. A mature plant can produce up to 26,000 seeds, some of which remain viable for almost 40 years. It is not eaten by cattle and quickly becomes a vigorous weed, replacing native plants and reducing the amount of pasture available for grazing.

The Oxeye Daisy is common more or less throughout the UK below about 600 m. It is a perennial plant, characteristic of grassland, that thrives in traditional hay meadows, overgrazed

Oxeye Daisy with Davies' Colletes *Colletes daviesanus* [PC] and Small Copper *Lycaena phlaeas* [PC]

pastures, field margins and roadside verges. It is also widely cultivated and a popular ornamental in gardens. And it features prominently in the meadows that are being created in gardens and parks, along motorway verges, and elsewhere in order to help our declining pollinators.

An Oxeye Daisy resembles a large common Daisy *Bellis perennis*. It grows to about 75 cm tall with white flowers 3–6 cm across. The flowering season is long, from May or June until September. The flowers attract diverse pollinators, mainly bees, hoverflies and butterflies.

Common Ragwort *Jacobaea vulgaris*

Common Ragwort is a native of northern Eurasia but occurs as an introduction in many other parts of the world, as far away as the USA, Argentina, Australia and New Zealand. It has been declared a noxious weed in much of Australia and New Zealand. Common Ragwort is a biennial or perennial and is abundant on disturbed land, overgrazed pastures and roadsides more or less throughout the UK. It is a tall erect plant, sometimes growing to more than a metre tall with dense clusters of yellow, daisy-like flowers. Its alternative names of Stinking Willie and Mare's Fart refer to the unpleasant smell of its foliage. There are other species of ragwort in the UK, all rather similar in general appearance. Hoary Ragwort *Jacobaea erucifolia*, Marsh Ragwort *Jacobaea aquatica* and the rare Fen Ragwort *Jacobaea paludosa* are natives. Several others are naturalised.

Common Ragwort is one of the five plants named in the Weeds Act 1959 that requires landowners to prevent the weeds from spreading. Common Ragwort is a particular concern to owners of horses because it contains toxic alkaloids. However, horses generally eat around any Common Ragwort growing in pastures and almost all cases of poisoning arise from animals eating contaminated hay or silage. The problem is greatly exacerbated by poor management of pastures. Overgrazing creates ideal conditions for Common Ragwort to establish itself and proliferate in pastures.

On a more positive note, Common Ragwort is an exceptionally useful plant for pollinators. It was rated seventh of the top ten plants for nectar production in a survey carried out by the AgriLand project funded by the UK Insect Pollinators Initiative. It has been reported that as many as 178 insect species use ragwort as a source of nectar and over 50 species (including their larvae) use it either as their exclusive food source, or a substantial part of their diet. The black and orange caterpillars of the Cinnabar moth *Tyria jacobaeae* are a conspicuous example. Their colours warn that they are unpalatable, having assimilated toxic alkaloids from the ragwort that are later passed on to the adult moths. Both caterpillars and moths are seldom eaten by predators.

Common Ragwort with Cinnabar moth *Tyria jacobaeae* caterpillars and Pearl Lacewing *Chrysopa perla*

Common Ragwort has a long flowering season from about mid-June to October. The composite flowers are 15–25 mm across with orange tube florets surrounded by yellow ray florets. A single plant can produce as many as 2,000 flowers in a season. The flowers are visited and pollinated by a good variety of small insects, including lacewings, male mosquitoes, hoverflies, blowflies, flesh flies, tachinid flies, numerous small solitary bees and many butterflies. The flowers often harbour thrips, both adults and larvae.

Coneflower or Sneezeweed (probably a *Helenium* cultivar)

Coneflowers in several genera (including *Echinacea*, *Rudbeckia* and *Helenium*) are native to North and South America. They are popular garden flowers and there are numerous cultivars. Coneflowers have a typical daisy-like inflorescence with a conspicuous cone-shaped head of dark brown or yellow tube flowers and yellow, orange or reddish turned-down, ray flowers. Varieties with a dark brown or blackish cone are often called Black-eyed Susans (as is also the climber *Thunbergia alata* in the Acanthaceae). And species of *Helenium* go by the common name Sneezeweed. Supposedly, native Americans made snuff out of the dried leaves and used the snuff to cause sneezing and thereby expel evil spirits from their head.

Most Coneflowers are very attractive to a great many different pollinators. We have not identified the variety of yellow coneflower that we have in our own garden but it is probably a *Helenium* cultivar. It has a cone of yellow tube flowers surrounded by rather short ray flowers and blooms late in the summer, from late July until late August or September. It is one of the best, all-round flowers for pollinators of all sorts, probably the best in our garden, attracting visitors of all shapes and sizes, including bumblebees, Honey Bees, numerous solitary bees, lacewings, hoverflies, small tachinid flies, satyrid (browns) and white butterflies, and a few small beetles. Our plants have even attracted two species of dark bees, Banded Dark Bee *Stelis punctulatissima* and Plain Dark Bee *Stelis phaeoptera*, that are rare in Norfolk.

Coneflower and Banded Dark Bee *Stelis punctulatissima*

Hemp-agrimony *Eupatorium cannabinum*

Hemp-agrimony (sometimes called Raspberries and Cream) is the only member of the genus that is native to Europe. Other species are found in North America and Asia. Hemp-agrimony is widespread in the UK, mainly below 350 m, common in the southern England and Wales but rarer, and mainly coastal, in Scotland. It also occurs in much of Europe, temperate Asia and locally in North Africa. It is an occasional garden escape in North America and other parts of the world. It is a robust perennial up to 1.5 m tall, found in damp woodland, on river banks, in marshes, fens and other wet habitats, and less often in dry habitats.

Hemp-agrimony has a reddish stem and three-lobed leaves that resemble those of Cannabis (hence its scientific name), although it is not related. The individual dioecious flowers are tiny, massed together in fluffy, dusty-pink clusters in a flat-topped inflorescence, up to 15 cm across. The flowers, which lack ray florets, have 5–6 pink tubular florets, each with an remarkably long, protruding style-stigma, surrounded by 8–10 purple-tipped bracts. The seeds are wind-dispersed, carried with the aid of a tuft of hairs.

Hemp-agrimony with Large Tiger Hoverfly *Helophilus trivittatus* [PC] **and Flesh Fly** *Sarcophaga carnaria*

Hemp-agrimony flowers from about July to September and is popular with a great variety of insects. It is particularly attractive to flies, including blowflies, flesh flies, tachinid flies and hoverflies. It also attracts nymphalid butterflies—Peacocks *Aglais io*, Small Tortoiseshells *Aglais urticae*, Red Admirals *Vanessa atalanta* and Painted Ladies *Vanessa cardui*. Visits by bees are rather scarce.

Adoxaceae (Elderberry family)

In older classifications this family was part of the honeysuckle family, Caprifoliaceae. It was separated by the Angiosperm Phylogeny Group, based on DNA evidence, and now consists of five genera and around 150 to 200 species. The three smallest genera—*Adoxa*, *Sinadoxa* and *Tetradoxa*—include just four species, all small herbaceous plants. Of these, Moschatel *Adoxa moschatellina* is widely distributed in the UK, Europe, Asia and North America. The other three species are found in China and Tibet. The two larger genera *Sambucus* and *Viburnum* are mostly woody and include the elders, Wayfaring-tree *Viburnum lantana* and Guelder-rose *Viburnum opulus*.

Elder or Elderberry *Sambucus nigra*

Sambucus is a genus of about 26 species, known as elders or elderberries. Most are shrubs or small trees found in many temperate or subtropical regions of the world. Elders are widely distributed in the Northern Hemisphere but restricted to just limited parts of South America and Australasia in the Southern Hemisphere.

Elder flowers [PC]

In the UK, the native Elder is common more or less throughout at altitudes below about 500 m. It grows prolifically in woodland, hedgerows and waste ground and reaches a height of about 15 m. Elder is often used for 'instant hedging' because stakes take root easily and grow quickly. Elder is popular with wild plant foragers because both its flowers and berries can be used to make elderberry wine, cordial or syrup. In various European countries, Elder flowers and berries are used to make sparkling elderflower 'champagne', brandy and liqueurs.

Elder flowers and berries, in the form of tea, juice, syrup or jelly, have long been used as dietary supplements and traditional medicines by herbalists and native European peoples. They have been used to treat an extraordinary range of disorders, including respiratory and gastrointestinal problems, skin rashes, and miscellaneous viral infections, including colds, influenza and fevers. However, it should be mentioned that there is little or no scientific evidence that any of these traditional 'cures' are effective.

Elder flowers are borne in flat-topped, umbel-like inflorescences, 10–30 cm across, that are in flower from about mid-May until mid-July. The individual flowers are small, white, fragrant and, like those of umbellifers, attractive to a wide range of small, generalist insect pollinators, particularly solitary bees, flies and beetles. In due course, the flowers produce drooping clusters of small purplish-black berries (technically drupes), 6–8 mm in diameter. The berries are ripe from about mid-August to early October and are consumed avidly by many birds, notably Blackbirds, Song Thrushes, Robins, Blackcaps, Garden Warblers and Starlings and, in small numbers, even by Jays and Magpies. Wood Pigeons, Bullfinches and Blue Tits also forage for elderberries but are seed predators, though Wood Pigeons probably disperse at least some seeds.

Caprifoliaceae (Honeysuckle family)

The family-level classification of the Caprifoliaceae has been controversial and undergone major revisions by the Angiosperm Phylogeny Group. Two familiar groups of species formerly placed in the Caprifoliaceae—the elderberries *Sambucus* and viburnums *Viburnum*—have been removed to the Adoxaceae, a revision also followed by Stace. However, two other former separate families—the valerians (Valerianaceae) and teasels (Dipsacaceae)—have been reduced to subfamilies by the Angiosperm Phylogeny Group and added to the Caprifoliaceae. This treatment has not been followed by Stace, who continues to recognise the Valerianaceae and Dipsacaceae as distinct families.

According to the APG's view of the Caprifoliaceae, the family now consists of 42 genera and between 850 and 900 species, with major centres of diversity in eastern North America and eastern Asia. As far as the UK is concerned, the family, as now defined, contains only the Honeysuckle *Lonicera periclymenum* and the Twinflower *Linnaea borealis*. In addition, there are numerous naturalised species and garden escapes, including the Fly Honeysuckle *Lonicera xylosteum*, which has been considered as possibly native in parts of the South Downs of West Sussex, though more likely it is an historical introduction. The family also includes many ornamental shrubs and vines that, in the UK, are more or less confined to gardens (e.g. species of *Abelia* and *Weigia*.

Honeysuckle *Lonicera periclymenum*

Lonicera, with about 180 species commonly known as honeysuckles, is the largest genus in the family Caprifoliaceae as it now stands. Honeysuckles are native to temperate regions of North America and Eurasia. The genus includes shrubs and many woody vines, some of which are intensely fragrant and popular as garden ornamentals. However, several species—notably the Japanese Honeysuckle *Lonicera japonica*—are now considered to be invasive weeds in the USA (and elsewhere) that smother other plants by climbing over them and shutting out their light. Some of these species have been introduced to the UK, have escaped from gardens and are now naturalised and invasive.

The Honeysuckle that is native to the UK, *Lonicera periclymenum*, is also found in much of Europe, Turkey and northern Africa. It is found more or less throughout the UK below 600 m. Growing to 7 m in height, it is a hardy, deciduous climber, twining clockwise, found in woodland, scrub and hedgerows, and is well able to survive in sunlight, semi-shade or deep shade. The White Admiral *Limenitis camilla* butterfly lays its eggs on the foliage.

The bisexual flowers, which are stalkless, are arranged in showy clusters and bloom from about June to September. The flowers are pale yellowish-cream, deepening to pale orange when pollinated, with crimson tints on their exterior. The corolla is a 40–50 mm long tube, split at the tip into upper and lower lips. The flowers have a sweet, heady fragrance, especially on warm summer evenings, that is hugely attractive to night-flying moths—the principal pollinators and said to be able to detect the Honeysuckle's scent at distances of up to a quarter of a mile. The flowers are also visited by long-tongued bees and hoverflies by day.

Clusters of red berries ripen in August and September and are eaten by thrushes, Robins, Blackcaps and Whitethroats. Bullfinches and tits predate the seeds.

Honeysuckle and a *Syrphus* hoverfly

Valerianaceae (Valerian family)

Though formerly treated as a family—the Valerianaceae—the family has been reduced to a subfamily of the Caprifoliaceae by the Angiosperm Phylogeny Group. However, Stace continues to recognise the Valerianaceae, a family that consists of about 350 species in seven genera. Most species are annual or perennial herbs, with foliage that has a rather disagreeable odour. The family is found mainly in north temperate regions and in the mountains of South America. It is absent from Australia. In the UK, the family includes four valerians *Valeriana*, two of them native; five cornsalads *Valerianella*, two of them native; and the introduced and naturalised Red Valerian.

A few species, notably common valerian, have been used as medicinal herbs for centuries, particularly for insomnia. In the UK, during World War II, it was used to relieve the stress of air raids.

Red Valerian *Centranthus ruber*

Red Valerian is a native of the Mediterranean region but was introduced to many other parts of the world, including northern Europe, the USA and Australia, and soon escaped from gardens and became naturalised in the wild. It was introduced to the UK as early as the 1600s and quickly spread to the countryside. It is now common throughout much of the UK but scarce or absent in northern areas. Red Valerian is very tolerant of dry, alkaline soil conditions and even tolerates the lime in mortar. For this reason, and because it has fluffy, wind-dispersed seeds, it is often seen growing on old walls, as well as hedgerows, road and railway cuttings, cliffs and other rocky habitats, especially near the sea.

Red Valerian is an attractive perennial plant with stems that grow to heights of 40–80 cm or more. The flowers are mostly pinkish-red but sometimes deep-red or white. The individual flowers are small but arranged in showy clusters that are particularly attractive to a good variety of insect pollinators. Red Valerian blooms profusely for six months or more, from late spring until October, sometimes later. It is an excellent garden plant except that its wind-dispersed seeds allow it to self-seed so efficiently that it is liable to become invasive unless managed. Even so, it is a very attractive and useful plant.

Red Valerian has a strong scent, sometimes described as fragrant, sometimes as rank or mousy, and attracts many pollinators. Butterflies love it and a good variety, including whites, Brimstones *Gonepteryx rhamni*, browns and nymphalids, are frequent visitors, as are Hummingbird Hawk-moths *Macroglossum stellatarum*. Silver Y *Autographa gamma* moths are common visitors in irruption years. Others include many hoverflies, a few bees and an occasional ruby-tail wasp *Chrysis* sp.

Red Valerian with a ruby-tail wasp *Chrysis* sp. and Small Tortoiseshell *Aglais urticae*

Dipsacaceae (Teasel and Scabious family)

Though currently placed in the Caprifoliaceae by the Angiosperm Phylogeny Group, Stace continues to recognise the Dipsacaceae as a distinct family containing 350 species of perennial or biennial herbs or shrubs in 11 genera. The family is native to temperate areas of Europe, Asia and Africa and many species have been introduced elsewhere. A few are now naturalised. In the UK, there are five native species and another five that are naturalised.

Wild Teasel *Dipsacus fullonum*

Wild Teasel is one of about 15 species of *Dipsacus*, a genus native to Eurasia and North Africa. Wild Teasel has been introduced to North America, southern Africa, Australia and New Zealand and is sometimes considered to be a noxious weed. It is widespread in the UK, characteristic of damp grassland, roadsides and waste ground, mainly on clay soils.

Wild Teasel is a biennial. It is distinctive mainly for its statuesque appearance, reaching heights of two metres or more, with prickly stems and spiky seed heads, about 5–10 cm long, that persist for most of the winter. The dried plant is a often used in flower arrangements. In the past, the comb-like, dried seed heads were used in the wool industry to raise the nap on fabrics by 'teasing' the fibres. Hence the name Teasel. The plant has pairs of large leaves on opposite sides of its main stem. The leaves completely surround the stem and form cup-like receptacles that fill with rain water, creating a barrier that prevents ants and other insects from climbing the stem and reaching the flowers. It has been shown that the presence of dead insects in the pools increases seed set, implying that the pools may become a source of nutrients, as they do in bromeliads and pitcher plants.

Wild Teasel and Garden Bumblebee *Bombus hortorum*

Wild Teasel is in flower from July to August or September. The oval inflorescence is an array of lilac flowers that open first in a band around its middle and then open sequentially both upwards and downwards, forming two bands. The individual flowers are about 12–15 mm long and are self-fertile, though cross pollination results in many more viable seeds. Wild Teasel is reputed to be very attractive to many potential pollinators, though in our garden in attracts mainly bumblebees and just a few hoverflies and butterflies.

Wild Teasel plants are said to produce an average of about 3,300 small seeds that mature in mid-autumn. The seeds provide an important winter food source for Goldfinches.

Field Scabious *Knautia arvensis*

Field Scabious is a hairy perennial that is native of Europe and temperate regions of Asia. It was introduced to gardens in North America and is now naturalised in some areas. It is widespread and common in most of the UK, particularly in the south and does best on dry, calcareous soils, especially where there is chalk bedrock. It is a plant of downs, meadows, field margins, grassy roadsides and waste ground.

The name 'scabious' comes from the plant's use as a folk remedy for scabies—a skin complaint caused by mites that results in a rash and intense itching. The whole plant is astringent and is used, mainly externally, to treat eczema, cuts and burns.

Field Scabious is a robust plant that produces clumps of large leaves and long, wiry flowering stems that reach up to 50–100 cm tall. It flowers more or less throughout the summer months, from June until September or even October, and is attractive to pollinators, making it an excellent choice for any wildlife garden or flower meadow. The flowers are borne on graceful, branching stems. Each is a flattish collection of bluish-lilac florets which together form a flower head 15–40 mm in diameter. Field Scabious now includes a range of cultivated garden varieties and colours, some of which have escaped to the wild. Field Scabious produces plentiful nectar-rich flowers for several months—flowers that are irresistibly attractive to bees, hoverflies, butterflies and many other insect pollinators. Two of the bees—the Large Scabious Mining Bee *Andrena hattorfiana* and Small Scabious Mining Bee *Andrena marginata*—specialise on scabious pollen and rarely collect anything else. Field Scabious and other scabious species are plants that are highly recommended for wildlife gardens

Field Scabious with Six-spot Burnet Moths *Zygaena filipendulae* and Large Scabious Mining Bee *Andrena hattorfiana* [PC]

Araliaceae (Ivy or Ginseng family)

The Araliaceae is a family of about 50 genera and 1,400 species, mostly trees and shrubs with a few climbers. The family is widely distributed but is most diverse in the humid tropics. The family is closely related to the Apiaceae (umbellifers) and the two families are merged by a minority of plant taxonomists. In the UK, there are two native members of the family—Common Ivy and Atlantic Ivy *Hedera hibernica*—though the two forms are regarded as a single species by some authorities. Only Common Ivy is considered here.

The family Araliaceae has limited economic importance. Ivies are grown as ornamentals and the Rice-paper Plant *Tetrapanax papyriferum* from Taiwan is the source of rice paper. Ginseng (various species of *Panax* is believed to have multiple medical benefits It is used, particularly in China, as a stimulant, to lower cholesterol levels, treat diabetes, reduce stress and manage sexual dysfunction in men.

Common Ivy flowers [PC]

Common Ivy *Hedera helix*

Common Ivy is a native species, tolerant of shade, and widely distributed in the UK. It is found in many habitats, including woodland, hedgerows, wasteland and on isolated trees and buildings where it climbs by means of sucker-like roots to heights of up to 30 m. It is sometimes accused of damaging or strangling trees but does not. It is not parasitic and has its own root system, one that is easily capable of obtaining all the nutrients and water that it requires. On the other hand, the weight and density of its evergreen foliage makes ivy-covered trees very vulnerable to being blown down or damaged in winter gales. Though Common Ivy is a shade tolerant species, it flowers more prolifically in bright sunlight.

Common Ivy is an exceptionally useful plant and should be valued for the important role that it plays in providing food for wildlife in the form of pollen and nectar for pollinators and berries for birds. It is especially valuable because it both flowers and fruits much later than most other plants. It flowers during September and October, sometimes as late as November, providing abundant pollen and nectar for any insects flying late in the year. The flower buds are also the preferred food plant for the second generation of caterpillars of the Holly Blue *Celastina argiolus* butterfly.

The numerous small, bisexual flowers of Common Ivy are pale green with yellow anthers and grouped in erect, rounded umbels. Pollen and nectar are freely exposed and provide an open house to a great diversity of small insects. In our own garden we used to have a luxuriant growth of ivy along 20 m of wall. It was rooted in the next door garden but sprawled over the wall onto our side. It attracted many flies, including greenbottles, flesh flies and huge Hornet Hoverflies *Volucella zonaria*. It flowered too late in the year for

Common Ivy and Hornet Hoverfly *Volucella zonaria*

Common Ivy with Red Admiral *Vanessa atalanta* and Comma *Polygonia c-album*

many species of butterflies but often attracted more than 20 Red Admirals *Vanessa atalanta* at a time, together with a few Small Tortoiseshells *Aglais urticae* and Commas *Polygonia c-album*. Red Admirals are often on the wing quite late in the year and ivy flowers are an important resource for them before they hibernate.

Unfortunately, our ivy was cut down in 2013, which was the year that the Ivy Bee *Colletes hederae*, an ivy specialist, was first recorded in Norfolk. Since then, the spread of the Ivy Bee has been rapid. It was first recorded in the UK, in Dorset, in 2001. It is now common and widespread in Norfolk, including our garden, and by 2016 had reached as far north as north Wales, the Lancashire/Cumbria border and Yorkshire.

Ripe ivy berries are available from about January to May, a time when most other fruits are already finished. Even though poisonous to humans, ivy berries are exceptionally nutritious with a high fat content and are consumed avidly by a wide size range of birds, including Robins, wintering Blackcaps, thrushes and Wood Pigeons. The latter are mainly seed predators rather than dispersers.

Common Ivy and Ivy Bee *Colletes hederae*

Apiaceae (Carrot and Parsley family)

The Apiaceae is a large family with about 3,700 species in over 450 genera. Thanks to the characteristic appearance of most species, it was the first flowering plant family to be recognised by botanists (in the late 16th century). The family is cosmopolitan, though most diverse and numerous in the northern temperate zone. The majority of species are herbs, many of them aromatic, and have a distinctive umbelliferous flower head. A few tropical species are shrubs or small trees, some reaching up to six or seven metres tall.

The Apiaceae have a remarkably wide range of culinary uses. Carrots and parsnips are major root crops and the Andean Arracacha *Arracacia xanthorrhiza* is an important subsistence crop in several South American countries, especially Peru and Bolivia. Celery, angelica, lovage and fennel are widely used as vegetables. Thanks to their aromatic nature, other umbellifers, notably chervil, culantro and parsley, are popular aromatic herbs, while spices derived from umbellifer seeds, many of which contain essential oils, include anise, caraway, coriander, cumin and dill.

However, not all umbellifers are edible so, given that the dangerously poisonous Hemlock and Hemlock Water-dropwort *Oenanthe crocata* resemble other umbellifers, it is important that wild plant foragers are very careful. The Greek philosopher, Socrates, was sentenced to execute himself by drinking a beverage of poison Hemlock. And Hemlock Water-dropwort has been used to stupefy fish. The latter is said to be the most poisonous British plant and to have been used in olden-days Sardinia to dispose of criminals and old people. Death is accompanied by a spasm of the facial muscles, resulting in an involuntary grimace—a 'sardonic grin'.

Including a few introductions, there are well over 50 umbellifers in the UK, many of them so superficially similar in appearance that they can be difficult to identify. Many species, including Cow Parsley *Anthriscus sylvestris*, Rough Chervil, Alexanders *Smyrnium olusatrum*, Hogweed and Wild Carrot, are common and familiar, often growing in dense stands along country roads and hedgerows. Numerous common umbellifers come into flower in succession throughout the spring and summer. For example, Alexanders and Cow Parsley flower early, from March or April until May or June; followed by Rough Chervil from May to July; Wild Carrot, Hogweed and Wild Angelica *Angelica sylvestris* from June to August or later; and Upright Hedge-parsley *Torilis japonica* and Hemlock in July and August.

The inflorescence is the most characteristic feature of the family Apiaceae. Most species are unmistakably 'umbellifers' with flowers arranged in showy, flat-topped, compound umbels. Individual flowers are small with five petals, most often white but sometimes pinkish or yellow. In the majority of species, flower stalks radiate from the tip of the stem, ending in a secondary umbel with rays that end in flowers. Most umbellifers are regarded as 'promiscuous' because their flowers provide easily accessible nectar that attracts a great diversity of small, unspecialised insects with short tongues. The best in our experience include Hogweed, Hemlock and Wild Carrot. Common visitors include hoverflies, tachinid flies, midges, small wasps, sawflies, small beetles and small solitary bees. Umbellifers attract very few bumblebees or other large bees, and equally few butterflies and moths. Nevertheless, umbellifers are doubly useful because they make good 'companion plants' that are attractive to the many predatory or parasitic flies, wasps and ladybirds that prey on or parasitise caterpillars and other larvae of garden pests. In spite of the high frequency of visits to umbellifers by potential pollinators, the close proximity of flowers on the same plant ensures that self-pollination is common.

Rough Chervil *Chaerophyllum temulum*

Rough Chervil is found throughout most of Europe, although it is rare around the Mediterranean. Its range also extends eastwards into western Asia, including Turkey and the Caucasus. Within the UK, the plant is common throughout most of England and Wales but local in Scotland, mainly in the south. Rough Chervil is a biennial or perennial, growing to about 1 m tall, and found in rough grassland, hedgerows, woodland edges, roadside verges and wasteland. It tolerates full sun but is found more often in dappled shade.

Unlike most other members of the genus *Chaerophyllum* (35 species found in Europe, Asia, northern Africa and North America), Rough Chervil contains various toxins. If handled, the sap can cause blistering, inflammation and rashes on skin and, if consumed, the plant causes gastro-intestinal

Rough Chervil and Black-and-yellow Longhorn Beetle *Rutpela maculata*

discomfort, symptoms of drunkenness and drowsiness. The Dutch physician, Herman Boerhaave (1668–1738), is said to have used a decoction of Rough Chervil and sarsaparilla to treat a case of leprosy. The treatment was successful though temporary blindness was a severe side effect.

Rough Chervil is rather similar to Cow Parsley *Anthriscus sylvestris* but is less competitive. It flowers a little later and can be distinguished by its hairy stem which rough to the touch and covered with purple spots and blotches. It flowers from late May until July and attracts a diverse range of small insect pollinators, including hoverflies, beetles and a few solitary bees.

Wild Carrot *Daucus carota*

Wild Carrot is native to temperate areas of Europe and nearby Asia and is also naturalised in parts of North America and Australia. In North America, the Wild Carrot is known as Queen Anne's Lace, partly because its dainty, creamy-white flower heads resemble lace. And also because Queen Anne, consort of James I, is said to have pricked her finger and spilled a drop of blood onto lace, making a pattern that resembled a Wild Carrot's single blood-red flower at the centre of a lacy white umbel.

The cultivated orange carrots that we eat nowadays are a cultivar of the Wild Carrot that probably originated in Afghanistan and was further developed by the Dutch in the mid 17th century. The Wild Carrot itself is only edible when very young and soon becomes too tough and woody to be worth eating. Wild plant foragers should be careful when handling the plant because skin contact with the foliage can result in significant skin irritation or even blistering. It is also said that medicinal properties of Wild Carrot seeds have been used by women for over 2,000 years to control fertility.

Wild Carrot is common throughout most of the UK, except for mountainous regions. It grows to about a metre tall and has finely divided leaves and umbels about 3–7 cm across. Several characters make identification relatively easy—it often has a red flower or flowers at the centre of its umbels; it has conspicuous forked bracts just below the umbels; and the developing seed heads fold inwards, making a distinctive 'bird's nest' structure. Wild Carrot is in flower from June until September. In our experience it attracts a particularly diverse selection of small insect pollinators, including solitary bees, sawflies and ichneumon wasps, tachinid and flesh flies, fungus gnats, lacewings and small beetles. In fact, Wild Carrot is a particularly good companion plant because it attracts many of the small wasps and flies that predate or parasitise many garden pests.

Wild Carrot [PC] and Flesh Fly *Sarcophaga carnaria*

Wild Parsnip *Pastinaca sativa*

Wild Parsnip is native to Eurasia where it has been used as a root vegetable for centuries. It was cultivated by the Romans and served as a sweetener until cane sugar became widely available in Europe. Wild Parsnip was introduced to North America at about the same time by both French colonists in Canada and by the British in their Colonies along the Atlantic coast. It is now widespread in the USA.

Wild Parsnips are quite closely related to carrots and, though sweeter, cultivated varieties are often used in similar ways. They can be baked, boiled, pureed, roasted, fried, grilled, or steamed and are particularly good in stews and casseroles. It is a nutritious plant, a good source of vitamins, antioxidants and minerals. Although a biennial, cultivated Wild Parsnips are usually grown as annuals and harvested late in their first winter. If left in the ground to mature during winter frosts, the long taproot develops a sweeter flavour. If unharvested, and left to complete its second year, the plant produces a flowering stem and seeds, but the taproot becomes woody and inedible.

Wild Parsnip is one of several umbellifers with stems and foliage that have to be handled with great care because they are protected by photo-sensitive chemicals that cause a severe skin rash or blistering, particularly in bright sunlight. The chemicals protect the plant from the Parsnip Moth *Depressaria radiella* caterpillars and other insect pests.

In the UK, Wild Parsnip is locally common in both England and Wales. It is a plant of scrub, rough grassland and roadside verges, mostly on chalk or limestone. It grows to a height of almost 2 m and is rather hairy with foliage that has a strong parsnip-smell when crushed. The umbels of yellow flowers are about 5–10 cm across and peak flowering is in July and August. Like most umbellifers, the flowers attract a diverse selection of mainly small generalist pollinators.

Wild Parsnip with Glass-winged Syrphus *Syrphus vitripennis* [PC]
and a Parsnip Moth *Depressaria radiella* caterpillar [PC]

Hogweed *Heracleum sphondylium*

Hogweed is native to most of Europe, including the UK, and its range extends eastwards to central Asia and south to parts of North Africa. In the UK, it is common throughout, except on high mountains. Hogweed gets its name from the distinctive, pig-like smell of its flowers.

Wild food foragers consider Hogweed, cooked like spinach or broccoli, to be one of the best tasting of any wild food plant available in the UK. Shoots can also be cooked in butter; young leaves can be used to flavour soups and stews; and the seeds can be used as a substitute for cardamom. The dead, hollow stems of Hogweed have also long been used by children as pea-shooters, water pistols and swords.

A word of warning. Contact with Hogweed sap, followed by exposure to sunlight, can cause allergic skin reactions. However, the reaction is nowhere near as severe as that caused by contact

Hogweed with Common Red Soldier Beetles *Rhagonycha fulva* [PC] and Yellow-bodied Black Fungus Gnat *Sciara hemerobioides*

with the related, but non-native, Giant Hogweed *Heracleum mantegazzianum* or some other umbellifers.

Hogweed is a robust biennial or perennial plant and is abundant along hedgerows, on roadside verges, waste ground and rough grassland. It grows to heights of 1.5–2 m, occasionally higher, and produces umbrella-like umbels of creamy-white or pinkish flowers, 10–15 cm across. Hogweed flowers from May or June until about September, sometimes even in winter. The flowers provide a great deal of nectar for pollinators. In fact, in a survey carried out by the AgriLand project, it was rated in the top 10 plants for nectar production (p. 35). Hogweed is attractive to a wide range of small generalist pollinators, including solitary bees, ichneumon wasps, flies and beetles.

Garden Parsley *Petroselinum crispum*

Garden Parsley is native to several countries in the Mediterranean region, including southern Spain, Portugal, Italy, Malta, and Greece as well as Morocco, Tunisia and Algeria in northern Africa. It is also naturalised, or a garden escape, elsewhere in Europe, including the UK, and very widely cultivated as a herb or vegetable in many parts of the world.

Garden Parsley is a widely used in cooking in just about all the areas in which it is native or has been naturalised. Many dishes are served

Garden Parsley and sawlies *Athalia circularis*

with chopped, fresh green parsley sprinkled on top. Two types of Garden Parsley are used—flat leaf and curly leaf versions. For a garnish, curly leaf parsley is usually preferred because it has a fresher look, especially when sprinkled on warm food. Root parsley, which looks like parsnip but tastes different, is yet another version of parsley, one that is popular in much of Europe, where it is used either as a snack or a vegetable in stews and casseroles.

Commonly planted in gardens and vegetable patches, Garden Parsley is a biennial that grows to about 60–75 cm tall. The umbels of small, greenish-yellow flowers appear in mid-summer from June until August. The flowers are attractive to numerous small, nectar-feeding insects, including small solitary bees *Andrena* and *Lasioglossum*, sawflies and hoverflies. Goldfinches feed on the seeds that appear later in the autumn.

Hemlock *Conium maculatum*

Hemlock is a biennial plant native to much of Europe and North Africa. It has also been introduced and naturalised elsewhere in the world, including parts of North and South America,

Asia, Australia and New Zealand. In the UK, Hemlock is an ancient introduction that is now common along roadsides or on waste ground and particularly profuse on damp ground alongside rivers and ditches. It occurs throughout most of England and Wales but is rare or absent in northern England, Scotland and Northern Ireland.

Hemlock is notorious for being very poisonous. Its toxins are alkaloids that cause muscular paralysis and respiratory failure. The Greek philosopher, Socrates, was famously tried, found guilty of 'impiety' and 'corrupting the young', and sentenced to execute himself by drinking a potion of Hemlock. Nowadays, cases of Hemlock poisoning are actually very rare, even among foragers for wild food plants, who routinely collect numerous edible umbellifers, such as Wild Carrot and Hogweed, that are superficially similar to Hemlock. The latter is easily distinguished by its hollow purple-spotted stems and unpleasant 'mousy' smell.

In appearance, Hemlock is a typical umbellifer that grows to 2–2.5 m tall with large, finely divided leaves and umbels of small white flowers. It is in flower from about June to August and, like most umbellifers, attracts small generalist pollinators, particularly flies, including many hoverflies.

Hemlock [PC] and Humming Syrphus *Syrphus ribesii*

Wild Angelica *Angelica sylvestris*

Wild Angelica is a short-lived perennial, native to Europe and central Asia, and introduced to North America, including Canada. It is common more or less throughout the UK, below about 850 m. It is found mainly in damp, lightly shaded habitats, including woodland, damp meadows, streamsides, ditches and fens.

Wild Angelica has often been used as a 'poor man's' substitute for Garden Angelica (next species). The traditional use for both Angelicas was for making candied decorations on cakes but Garden Angelica was preferred because its stems are less tough and bitter than those of Wild Angelica. Wild Angelica

Wild Angelica [PC]

Wild Angelica with Pale-saddled Leucozona *Leucozona glaucia* hoverfly [PC] and Plantain Wasp-sawfly *Tenthredo omissa* [PC]

was also used as a herbal medicine, useful for indigestion, minor stomach upsets, lung and chest problems, rheumatism and even corns. Nevertheless, when available, Garden Angelica was usually preferred.

A robust, handsome plant, Wild Angelica reaches heights of up 2.5 m. Its stems are hollow, suffused purplish and have distinctive inflated sheaths where leaf and flower stalks join the stems. The large, umbrella-like, rounded umbels are 3–15 cm across with tiny flowers that are white or tinged pink and in flower from July to September or October. The flowers are attractive to many small pollinating insects, particularly hoverflies, ichneumon wasps and small solitary bees. The seeds have thin, papery wings and are wind-dispersed.

Garden Angelica *Angelica archangelica*

Garden Angelica is a biennial plant native to Russia, Scandinavia, Greenland, Iceland and the Faroe Islands. It is naturalised in the UK where it is found in damp soil on riverbanks or close to ponds, mainly in the south.

Garden Angelica is an aromatic plant that has been cultivated since the 10th century as a medicinal plant, as a vegetable and as a source of flavouring, especially for liqueurs. As a medicinal plant, garden angelica has been used to treat heartburn, flatulence, arthritis, circulation problems, respiratory catarrh, plague, and insomnia. However, it has also been known for its culinary use as a flavouring for omelettes and fish. And its edible stems are candied in sugar syrup, coloured green and

Garden Angelica and Furry Dronefly *Eristalis intricarius*

used as cake decoration. It is also much used to flavour liqueurs, spirits and fortified wines, including Chartreuse, Benedictine, gin, vermouth and Dubonnet.

As a biennial, Garden Angelica grows only leaves in its first year but, in its second, it grows into an impressive, statuesque plant with a fluted stem, reaching a height of 2 m or more. The large globular umbels bear numerous small, yellowish or greenish, flowers that bloom in July. The flowers are very attractive to smallish pollinating insects, particularly flies and small bees.

MONOCOTYLEDONS

Iridaceae (Iris family)

The iris family has a cosmopolitan distribution and contains about 70 genera and over 2,000 species, most of them herbaceous with just a few shrubs. Species of Iridaceae are found in temperate, subtropical and tropical regions but occur most commonly in seasonally dry regions. They are particularly common and diverse in southern Africa. The family is well known for numerous ornamental genera that are grown commercially, such as irises, crocuses, freesias and gladioli. The Saffron Crocus *Crocus sativus* is the source of the spice saffron—the most valuable spice by weight. The important parts of the flower are the crimson, 3-pronged styles and their stigmas, called threads, which are collected and dried for use as a food colouring agent and seasoning. As many as 75,000 plants are needed to produce a pound of saffron which can cost as much as US $5,000 per kg or more. The Saffron Crocus is unknown in the wild but has been cultivated for more than 3,500 years, nowadays mainly in the Mediterranean region and eastwards through Iran (which has up to 90% of total world production) to Kashmir.

Plants in the Iridaceae are perennials that have underground storage organs in the form of bulbs, corms or rhizomes. Many species (not all) are adapted to seasonal climates that have a pronounced dry or cold season that is unfavourable for plant growth, during which time above-ground stems and leaves die back and the plants become dormant. For example, by retreating underground in this way, the many species of Iridaceae that grow in African fynbos, veld or grassland can even survive dry season fires. The fires are actually beneficial in that they burn competing vegetation and fertilise the soil with ash. With the arrival of the first rains, the dormant bulbs or corms undergo a rapid spurt of growth, producing flowers before they have too much competition from the regrowth of other vegetation.

Yellow Iris *Iris pseudacorus*

The Yellow Iris, or Yellow Flag, is native to much of Europe, western Asia and north-western Africa. In the UK it can be found almost throughout in wet habitats below about 600 m. In western Scotland, including the Western Isles, extensive stands of Yellow Iris provide an important feeding and breeding habitat for rare and endangered Corncrakes.

The Yellow Iris is a perennial that grows to heights of about 100–150 cm sometimes higher. It likes very wet conditions, including wet meadows, muddy shores and flooded wetlands, swamps and fens where its roots are often submerged. It spreads rapidly by means of both rhizomes and

Yellow Iris and Common Carder Bumblebee *Bombus pascuorum*

its water-dispersed seeds and it can become invasive if it spreads too much and takes over a small pond or other important habitat. Its seeds, which float in water, ensure that it readily colonises new areas of suitable habitat. In fact, it has the potential to become an invasive species and outcompete other aquatic plants. On the other hand, it has been used for sewage treatment and also as a water purification treatment because it has the ability to take up heavy metals through its roots.

The large, bright yellow flowers of the Yellow Iris are in bloom from May to August. Their characteristic structure is thought to be the inspiration for the fleur-de-lis—an enduring symbol of France and French nobility. Irises have an unusual structure. In effect, an iris has become three separate functional flowers, each of which can be pollinated individually. The Yellow Iris secretes copious nectar and was rated second in daily nectar production per flower in a UK survey conducted by the AgriLand project. Bumblebees and long-tongued flies are the main visitors and pollinate most of the flowers.

Amaryllidaceae (Daffodil and Onion family)
In the past, the family was sometimes combined with the Liliaceae. However, the most recent classification, based on DNA evidence, includes about 75 genera and 1,600 species, allocated to three subfamilies: Amaryllidoideae including amaryllis, daffodils and snowdrops; Agapanthoideae with only agapanthus; and Allioideae including onions, leeks, garlic and chives. Plants in the latter subfamily produce allyl sulphide compounds which give them their characteristic and challenging smell.

The family includes numerous popular, ornamental plants, such as daffodils, snowdrops, amaryllis and ornamental alliums. It also includes vegetables, such as onions, leeks and garlic. Here we are concerned only with one of the ornamental onions *Allium hollandicum*. Other members of the family that grow in our garden—early flowering snowdrops and daffodils—attract virtually no pollinators, probably because the weather in early spring is now so unpredictable and extreme (see Chapter 7. Global warming, p. 24)

Persian Onion or Dutch Garlic *Allium hollandicum* 'Purple Sensation'
The Persian Onion is a perennial plant that is native to Iran and Kyrgyzstan in the Middle East. Estimates of the number of *Allium* species vary widely, ranging from as low as 260 to almost 1,000. The vast majority are found in temperate areas of the Northern Hemisphere with the greatest diversity being found in the eastern Mediterranean.

Persian Onion with Peacock *Aglais io* and Four-banded Flower Bee *Anthophora quadrimaculata* [PC]

Many species of *Allium* have been cultivated as ornamentals. The variety that is dealt with here—*Allium hollandicum* 'Purple Sensation'—is a particularly popular hybrid that has been given the Royal Horticultural Society's Award of Garden Merit. 'Purple Sensation' is a bulb-forming plant that has spherical umbels (about 10 cm in diameter) of attractive purple, star-shaped flowers atop a leafless stem up to 90 cm tall. The flowers have a strong smell of onions or garlic.

The flowers bloom in early summer in late May and June and produce a lot of nectar that attracts a good variety of insect pollinators. Butterflies of several species, including Peacocks *Aglais io*, Small Tortoiseshells *Aglais urticae* and 'whites', are the most frequent visitors.

Asparagaceae (Asparagus and Bluebell family)

As conceived by the Angiosperm Phylogeny Group III (2009), based on phylogenetic studies, the Asparagaceae now lumps together seven subfamilies (all formerly families) that together comprise 114 genera and about 2,900 known species. With a near worldwide distribution, the family is now so diverse that it is difficult to see any obvious morphological similarities that might indicate close relationships. In fact, the family members are united primarily by genetic and evolutionary relationships based on DNA sequencing, rather than traditional botanical systematics. Of the seven subfamilies involved, the only one to concern us here is the Scilloideae, formerly the Hyacinthaceae, which includes both the Bluebell and grape-hyacinths *Muscari*.

Bluebell *Hyacinthoides non-scripta*

The Bluebell, sometimes called the English Bluebell, is native to the western Atlantic fringe of Europe. It is also naturalised in Germany and Italy and has been introduced into parts of North America. The UK is a bluebell stronghold where the species is associated with ancient woodland. Indeed, Bluebell woods are one of the most distinctively British of all our plant communities. Many thousands of bulbs are found in some woodlands, where they create a glorious blue carpet that rivals any of the most dazzling wildflower spectacles in the world. It is no surprise that the Bluebell is often regarded as the United Kingdom's favourite flower.

In the past, Bluebells were used in a number of curious ways. Their sticky sap was used to both bind the pages of books and glue the feathers onto arrows. And, in Elizabethan times, starch made from crushed bulbs was used to stiffen the ruffs characteristic of the age. There are also many folklore tales about

Bluebell

Bluebell woods that involve dark fairy magic and enchantments. Perhaps the darkest magic is the belief that hearing a Bluebell's bells ringing attracts a bad fairy and your death soon after. Bluebells are too toxic to have been widely used in herbal medicine, though they were sometimes used as a diuretic or styptic (to control bleeding).

Nowadays, the English Bluebell is seriously threatened by habitat destruction, the illegal collection of wild bulbs, and by the Spanish Bluebell *Hyacinthoides hispanica*, introduced into British gardens in the 17th century. English and Spanish Bluebells readily hybridise and, unfortunately, the resulting hybrids can back-cross with either parent. The hybrids are very vigorous and, if interbreeding is allowed to continue, there is a distinct possibility that our native English Bluebell could be out-competed, swamped and become extinct.

While the English Bluebell is still common in most of the UK, it is now protected under the Wildlife and Countryside Act (1981 and 1998). Landowners are prohibited from removing common Bluebells on their land for sale and it is a criminal offence to remove the bulbs of wild common Bluebells. This legislation was strengthened in 1998 under Schedule 8 of the Act making any trade in wild common Bluebell bulbs or seeds an offence, punishable by fines of up to £5,000 per bulb.

The Bluebell is common in woodland throughout most of lowland Britain, below about 600 m. As well as woodland, Bluebells grow along shady hedgerows and banks. The fragrant flowers are violet-blue in colour and bell-shaped with six petals with up-turned tips. Up to 20 flowers form an inflorescence, all characteristically hanging to one side of the stem. Bluebells flower from late April to June. They produce plentiful nectar and pollen, and attract long-tongued bumblebees and hoverflies, bee-flies and a few woodland butterflies. Honey Bees steal nectar by biting a hole in a flower to reach its nectary.

Garden Grape-hyacinth *Muscari armeniacum*

Muscari is a genus of perennial plants that are native to southern and central Europe, northern Africa and much of temperate Asia. One species—*Muscari neglectum*—is a rare native to breckland in East Anglia and naturalised elsewhere. Grape-hyacinths are popular ornamental garden plants and are now widely naturalised in northern Europe and North America. The plants produce dense spikes of blue, rounded flowers that supposedly resemble bunches of grapes—hence the common name.

Many species of grape-hyacinths can be difficult to identify. The present species—*Muscari armeniacum*—is widespread in the eastern Mediterranean region, from Greece and Turkey to Armenia. It has been commonly planted in European gardens since 1871 and is commonly grown in the UK. As a garden throw-out, it quickly becomes naturalised and spreads easily, usually in the vicinity of country villages and houses.

Grape-hyacinths, including this species, tend to bloom earlier in spring, during March and April, than many flowers, making them an important source of nectar for early emerging insects. The flowering spikes are 15–30 cm tall and the individual flowers are near spherical or urn-shaped, 4–7 m long, and constricted at one end to form an entrance surrounded by small white 'teeth'. The flowers secrete copious nectar and are extremely attractive to long-tongued insect pollinators, notably Hairy-footed Flower Bees *Anthophora plumipes* and Dark-edged Bee-flies *Bombylius major*.

Garden Grape-hyacinth and Dark-edged Bee-fly
Bombylius major

Commelinaceae (Spiderwort family)

The family Commelinaceae includes about 50 genera and 700 species of herbs, most of them with succulent stems and sheathing leaves. The family is well represented and diverse in the tropics of both the Old and New Worlds. Many species are cultivated as garden ornamentals or house plants, with favourites including the dayflowers *Commelina* and spiderworts *Tradescantia*.

The flowers are very variable in structure. They can be either radially symmetrical (actinomorphic) as in *Tradescantia* or bilaterally symmetrical (zygomorphic) as in *Commelina* and *Aneilema*. And they can be either bisexual or more rarely unisexual. The flowers are ephemeral, sometimes lasting for less than a day. They lack nectar so pollen is the only reward on offer to potential pollinators. Some flowers also have adaptations intended to deceive pollinators by appearing to offer larger rewards than are actually present. The adaptations take the form of yellow hairs that mimic pollen or staminodes that resemble stamens but lack pollen. In practice, bees and a few flies are the principal pollinators.

Virginia Spiderwort *Tradescantia virginiana*

There are about 70 species of spiderwort in the genus *Tradescantia*. All are herbaceous perennials, native to the New World, ranging from Canada to northern Argentina. Spiderworts were introduced into Europe in 1629 and are now popular ornamentals in many parts of the world.

The Virginia Spiderwort, which is the type species of *Tradescantia*, grows to about 60–70 cm in height. The beautiful violet-blue flowers have three petals, accented by six stamens with contrasting yellow anthers. Individual flowers remain open for less than a day but new ones open daily for up to six weeks.

The flowers have no scent and lack nectar, so pollen is the only reward. In North America the most important pollinators are said to be bumblebees and solitary bees. Other visitors include short-tongued hoverflies that feed on pollen but are probably ineffective as pollinators. The flowers in our garden attract short-tongued Marmalade Hoverflies *Episyrphus balteatus* and not much else.

Virginia Spiderwort and Marmalade Hoverfly
Episyrphus balteatus

Glossary

abiotic pollination—Pollination involving wind or water, not an animal.

actinomorphic—Having radial symmetry.

angiosperm—A plant that produces seeds inside flowers.

anther—The part of the stamen that manufactures and releases pollen.

apomixis—Asexual reproduction in which seeds are produced from unfertilised ovules.

bisexual—A flower with both male and female reproductive structures, including stamens and an ovary; the same as perfect or hermaphrodite.

bract—A modified leaf at the base of a flower, usually with a protective or advertising function.

buzz-pollination—Pollination in which a bee clings to a flower and vibrates its indirect flight muscles, causing pollen to be released from the anthers.

calyx—A collective name for the sepals.

carpel—A component of the female reproductive system, made up of an ovary, style, and stigma; one or more carpels form a pistil.

cleptoparasite (or kleptoparasite)—An animal that steals food, including stored food, from another.

clone—Here refers to a population of plants produced asexually and genetically identical to their common ancestor.

corolla—A collective name for the petals.

cross-pollination—Pollen being moved from the stamens of one flower to the stigma of a genetically distinct flower of the same species.

deceit pollination—The attraction of pollinators without a genuine reward.

dicots—Short for dicotyledons; one of two groups of flowering plants, this one being characterised by having two embryonic seed leaves at germination (see monocots).

dioecious—Having male and female flowers on separate plants (see monoecious).

elaiophore—An oil or lipid producing gland in a plant.

gigatonne—1 billion metric tons.

greenhouse gases—Any gases emitted by human activities that absorb and emit heat energy and so contribute to the greenhouse effect. Carbon dioxide, methane and nitrous oxide are the most important greenhouse gases.

gymnosperm—A seed plant lacking flowers e.g. a conifer, cycad, etc.

heliotropism—Diurnal or seasonal motion of a flower so as to face the sun.

heterostyly—Having two or three different style lengths in different individual flowers.

inbreeding—Reproduction between genetically related individuals.

inflorescence—A cluster of flowers, usually forming a functional unit in terms of the attraction of pollinators.

metric ton or tonne—1,000 kilograms.

monocots—Short for monocotyledons; one of two groups of flowering plants, this one being characterised by having a single seed leaf at germination (see dicots).

monoecious—Having separate male and female flowers on the same plant (see dioecious).

monolectic—Refers to a bee that collects pollen from the flowers of a single plant species.

Mullerian mimicry—mimicry in which two or more unpalatable animals have evolved similar warning colours and behaviour as shared protection against predators.

nectar—A watery secretion containing sugars and other nutrients.

nectar guide—A contrasting pattern on flower petals that indicate the location of nectaries.

nectary—A structure or tissue that secretes nectar.

oligolectic—Refers to a bee that specialises in collecting pollen from one genus or species (or a few genera or species) of flowering plants.

outbreeding—Breeding between individuals that are not closely related.

ovary—Female organ at the base of the pistil that contains the ovules.

ovoviviparous—An animal that produces young by means of eggs which are hatched within the body of the parent.

ovule—Female organ that becomes a seed after fertilisation.

parasitoid—An organism whose larvae live as parasites which eventually kill their hosts.

perfect—See bisexual.

perianth—Collective term for the calyx and corolla.

petal—A sterile segment of the corolla, often coloured or scented.

pheromone—A volatile chemical released by one organism that influences the behaviour of another.

pistil—A structure formed when a flower's carpels are fused together to resemble a single, giant carpel.

pollinarium—The pollen-bearing structure of orchids that is attached to insects during pollination; composed of a viscidium, stipe, and pollinia.

pollination syndrome—A suite of flower characters that match the characteristics of appropriate pollinators.

pollinium—An aggregate mass of pollen grains that are transferred during pollination as a single unit; characteristic of orchids and milkweeds.

polylectic— Refers to a bee that collects pollen from the flowers of a variety of different plants.

protandrous—Refers to a flower in which the stamens release pollen and wither before the stigma is receptive.

protogynous—Refers to a flower in which the stigma is receptive before pollen is released.

pseudocopulation—A form of pollination brought about by deceit; male insects attempt to mate with a flower that resembles a female of the species in some important way, visually or by odour.

pyrrolizidine alkaloid—Toxic protective chemical in plants that is collected and used by specialised insects for their own defence.

scopa—A storage area for the pollen that bees groom from their body and carry to their nest; it can be on their legs, thorax, or under the abdomen.

self-pollination—Pollination with pollen from the same flower or same plant.

sepal—A segment of the calyx.

sonication—See buzz-pollination.

spadix—A spike-like inflorescence of small flowers on a fleshy stem e.g. the inflorescence of aroids.

spathe—A sheath-like bract that the surrounds a spadix.

stamen—A male flower organ composed of a filament and pollen-bearing anther.

staminode—A sterile stamen, often visually attractive, that provides a food reward to potential pollinators.

stigma—The terminal portion of the pistil where pollen is received.

stigmatic secretion—Sticky fluid on the stigma that is involved in pollen reception; sometimes becomes a reward for small pollinating insects.

style—The slender stalk that connects the ovary to the stigma.

tepal—Refers to sepals and petals that resemble each other.

tonne—A metric ton or 1,000 kilograms, equal to approximately 2,204 pounds.

trapliner—In relation to bees, moths or butterflies, a species or individual that regularly visits widely dispersed flowers along a specific route.

umbel—A flattish-topped inflorescence, the flowers on many short stalks emanating from a central point.

zygomorphic—Having bilateral symmetry.

Further reading

Ball, S & Morris, R. 2013. *Britain's Hoverflies. A field guide.* Princeton University Press, Woodstock, U.K.

Bernhardt, P. 1999. *The Rose's Kiss: a Natural History of Flowers.* Island Press, Washington, D.C.

Buchmann, S. L. 2015. *The Reason for Flowers.* Scribner, New York. *Climate change.* 2020.

Doyle, T. *et al.* 2020. Pollination by hoverflies in the Anthropocene. *Proc. R. Soc. B* 287: 20200508. http://dx.doi.org/10.1098/rspb.2020.0508

Falk, S. 2015. *Field Guide to the Bees of Great Britain and Ireland.* Bloomsbury, London.

Fogden, M.P.L. & Fogden, P.M. 2018. *The Natural History of Flowers.* Texas A&M University Press, College Station, Texas.

Garibaldi, L.A. *et al.* 2013. Wild Pollinators Enhance Fruit Set of Crops Regardless of Honey Bee Abundance. *Science* 339 (6127): 1608-11. doi:10.1126/science.1230200.

Geldmann, J. & Gonzalez-Varo, J.P. 2018. Conserving honey bees does not help wildlife. *Science* 359 (6374), 392–393. doi: 10.1126/science.aar2269.

Goody, J. 1993. *The Culture of Flowers.* Cambridge University Press, Cambridge.

Hallmann C.A. *et al.* 2017. More than 75 percent decline over 27 years in total flying insect biomass in protected areas. *PLOS ONE* 12(10): e0185809. https://doi.org/10.1371/journal.pone.0185809

Jones, L., Brennan, G.L., Lowe, A. *et al.* Shifts in honeybee foraging reveal historical changes in floral resources. *Commun Biol* 4, 37 (2021). https://doi.org/10.1038/s42003-020-01562-4

Jones, M.W. *et al.* 2020. Climate Change Increases the Risk of Wildfires. https://sciencebrief.org/briefs/wildfires

Lenton, T.M. *et al.* 2019. Climate tipping points—too risky to bet against. *Nature* 575, 592–595. doi: 10.1038/d41586-019-03595-0.

Meeuse, B. & Morris, S. 1984. *The Sex Life of Flowers.* Faber and Faber, London.

Neukom, R. *et al.* 2019. No evidence for globally coherent warm and cold periods over the preindustrial Common Era. *Nature* 571, 550–554. https://doi.org/10.1038/s41586-019-1401-2.

Orr, M. *et al.* 2020. Global patterns and drivers of bee distribution. *Current Biology.* https://doi.org/ 10.1016/j.cub.2020.10.053

Owens, N. 2017. *The Bees of Norfolk.* Pisces Publications, Newbury, U.K.

Owens, N. 2020. *The Bumblebee Book. A guide to Britain & Ireland's bumblebees.* Pisces Publications, Newbury, U.K.

Paschalidou, F.G., *et al.* 2020. Bumble bees damage plant leaves and accelerate flower production when pollen is scarce. *Science* 368: 881. doi:10.1126/science.aaz7435.

Powney, G.D., *et al.* 2019. Widespread losses of pollinating insects in Britain. *Nature Communications* 10: 1–6.

Proctor, M., Yeo, P. & Lack, A. 1996. *The Natural History of Pollination.* Timber Press, Portland, Oregon.

Sánchez-Bayo, F. & Wyckhuys, K.A.G. 2019. Worldwide decline of the entomofauna: A review of its drivers. *Biological Conservation* 232: 8–27.

Sasgen, I., Wouters, B., Gardner, A.S. *et al.* 2020. Return to rapid ice loss in Greenland and record loss in 2019 detected by the GRACE-FO satellites. *Commun Earth Environ* 1, 8. https://doi.org/10.1038/s43247-020-0010-1

Siegert, M. *et al.* 2020. The Arctic and the UK: climate, research and engagement, Grantham Institute Discussion Paper 7, 8pp, Imperial College London. Doi: https://doi.org/10.25561/80095.

Soroye, P., Newbold, T. & Kerr, J. 2020. Climate change contributes to widespread declines among bumble bees across continents. *Science* 367: 685–688

Stace, C. 2019. *New Flora of the British Isles.* 4th edition. Cambridge University Press, Cambridge.

Thomas, J.A. & Lewington, R. 1991. *The Butterflies of Britain and Ireland.* Dorling Kindersley.

Tiusanen, M. *et al.* 2016. One fly to rule them all—muscid flies are the key pollinators in the Arctic. *Proceedings of the Royal Society B: Biological Sciences.* doi: 10.1098/rspb.2016.1271.

Varah, A. *et al.* 2020. Temperate agroforestry systems provide greater pollination service than monoculture. *Agriculture, Ecosystems & Environment.* doi: 10.1016/j.agee.2020.107031.

Walton, R.E., Sayer, C.D., Bennion, H. & Axmacher, J.C. 2020. Nocturnal pollinators strongly contribute to pollen transport of wild flowers in an agricultural landscape. *Biol. Lett.* 16: 20190877. http://dx.doi.org/10.1098/rsbl.2019.0877.

Wang, J., Feng, L., Palmer, P.I. *et al.* 2020. Large Chinese land carbon sink estimated from atmospheric carbon dioxide data. *Nature* 586: 720–723 https://doi.org/10.1038/s41586-020-2849-9.

Willmer, P. 2011. *Pollination and Floral Ecology.* Princeton University Press, Princeton, New Jersey.

Wotton, K.R. *et al.* 2019. Mass Seasonal Migrations of Hoverflies Provide Extensive Pollination and Crop Protection Services. *Current Biology* 29: 2167–2173. http://doi.org/10.1016/j.cub.2019.05.036.

Index

Number in **bold** refer to the page numbers of recommended plants for pollinators listed in chapter 15.